Serotonin - Neurotransmitter and Hormone of Brain, Bowels and Blood

Edited by Kaneez Fatima-Shad

Published in London, United Kingdom

Serotonin - Neurotransmitter and Hormone of Brain, Bowels and Blood
http://dx.doi.org/10.5772/intechopen.1000435
Edited by Kaneez Fatima-Shad

Contributors
Alfredo Briones-Aranda, Ammarah Baig, Citlali Trueta, Diana Casique-Aguirre, Javeria Tanveer, Josselin Corzo-Gómez, Kaneez Fatima-Shad, Kanza Khan, Kiran Bala, Mark W. Morningstar, Mauricio Megchún-Hernández, Montserrat G. Cercós, Mushtaq Margoob, Neelofer Jan, Nicole Spiegelaar, Nosheen Bashir, Rukhsana Rubeen, Sebastian Warma, Shahana Rasheed Qureshi, Shazia Kouser

First published in London, United Kingdom, 2024 by IntechOpen
IntechOpen is the global imprint of INTECHOPEN LIMITED, registered in England and Wales, registration number: 11086078, 167-169 Great Portland Street, London, W1W 5PF, United Kingdom

British Library Cataloguing-in-Publication Data
A catalogue record for this book is available from the British Library

Additional hard and PDF copies can be obtained from orders@intechopen.com

Serotonin - Neurotransmitter and Hormone of Brain, Bowels and Blood
Edited by Kaneez Fatima-Shad
p. cm.
Print ISBN 978-0-85466-278-4
Online ISBN 978-0-85466-277-7
eBook (PDF) ISBN 978-0-85466-279-1

For EU product safety concerns:
IN TECH d.o.o., Prolaz Marije Krucifikse Kozulić 3, 51000 Rijeka, Croatia,
info@intechopen.com or visit our website at intechopen.com.

Meet the editor

Professor Kaneez Fatima Shad, an Australian neuroscientist with a medical background, earned her Ph.D. in 1994 from the Faculty of Medicine at UNSW, Australia. Following her doctoral studies, she pursued a postdoctoral fellowship at Allegheny University of Health Sciences in Philadelphia, USA. With a rich academic background, Professor Shad has shared her expertise in medical and biological sciences across various esteemed institutions worldwide, spanning Australia, the USA, UAE, Bahrain, Pakistan, and Brunei. Throughout her career, she has been actively involved in cutting-edge research, securing both local and international grants totaling over US$2.5 million. Her research endeavors have led to the development of innovative products, including a rapid diagnostic test for stroke and other vascular disorders such as schizophrenia. Professor Shad's contributions to academia extend beyond research, with over 76 articles published in refereed journals, nine books edited, and 10 book chapters authored. She is also a seasoned presenter, having shared her insights at over 100 international conferences. Furthermore, Professor Shad has played a crucial role in nurturing the next generation of scholars, having mentored 34 postgraduate students. Known for her dedication to education and research, Professor Shad serves as a mentor to students and a specialist in protocol development. Her expertise and passion for advancing knowledge in the field of neuroscience continue to inspire and shape the future of medical science.

Contents

Preface

Serotonin is the most ancient neurotransmitter found throughout the body. Serotonin and its metabolite 5-hydroxyindolacetic acid are present in different ratios in the brain and blood, indicating the difference in its metabolism.

5-hydroxytryptamine, also known as serotonin, is the major source of the regulation of our mood and is also responsible for most of the physiological and pathological processes within the human body.

This amazing molecule is not only involved in blood circulation but also involved in hemostasis and vasoconstriction. The enterochromaffin cells in the gut mucosa are the primary source of serotonin synthesis, which is partially regulated by the gut bacteria.

Serotonin is released in the blood; it is stored in the platelets and controls the balance between blood flow and clotting. Conversely, in the central nervous system, serotonin regulates functions from mood modulation to cognition and acts as a molecule of happiness.

Various regions of the brain have different ratios of serotonin and its metabolic end product 5-hydroxyindolacetic acid (5-HIAA), showing different metabolic rates of serotonin. Literature indicates that the highest serotonin metabolism occurs in the frontal neocortex, followed by the hippocampus and platelets, indicating that the demand for serotonin is highest in the frontal neocortex as this molecule is involved in behaviors from attention to appetite and aggressiveness.

I would like to thank Publishing Process Manager Nina Miocevic and her team for their support and guidance throughout my journey of editing this book.

Finally, I would like to emphasize that this book caters to the tastes of anyone who loves to understand why we eat, sleep, and seek happiness.

Kaneez Fatima-Shad
Professor of Neuroscience,
School of Life Sciences,
Faculty of Science,
University of Technology Sydney,
Sydney, Australia

Section 1

Biomarkers

Chapter 1

Insulin Impairment Disrupts Central Serotonin Synthesis: Implications for Stress Resilience

Nicole Spiegelaar and Sebastian Warma

Abstract

This chapter reviews the important neurophysiological mechanisms that drive symptoms characteristic of comorbid depression and metabolic disease. It outlines how insulin impairment in the periphery[1] interferes with central 5-hydroxyindole metabolism and ultimately restricts central[2] serotonin synthesis. More specifically, peripheral insulin impairment disrupts i) peripheral and central tryptophan stores, ii) tryptophan uptake into the brain, and iii) tryptophan hydroxylase-2 function. Central serotonin availability appears to be increasingly restricted by higher degree and duration of insulin impairment, which can lead to both physiological and behavioral positive feedback loops experienced by individuals as a spiral of deteriorating mental health and tryptophan metabolism. Serotonin and its metabolites are fundamentally homeostatic regulators that serve to enhance adaptive response to stress in all organisms. Considering this essential trait, this review proposes that: disruptions in normal 5-hydroxyindole metabolism of tryptophan during impaired insulin function will disrupt homeostatic adaptive capacity of central serotonin, thereby increasing vulnerability to emotional and energy disturbances, and limiting recovery from such disturbances.

Keywords: serotonin, insulin, tryptophan, diabetes, depression, complex adaptive systems, psychological resilience

1. Introduction

Serotonin (5-hydroxytryptamine or 5-HT) is an important neurotransmitter and hormone involved in various functions, such as emotion, cognition, learning, metabolism, sleep, platelet function, and gastrointestinal motility [1–5]. These functions depend on tight regulation of available 5-HT and other indole metabolites of Tryptophan (Trp) throughout the body and brain [1]. Research continues to uncover how imbalances of these molecules is associated with insulin impairment, and a cascade of disruption via complex interactions between the mind and body. This review highlights bidirectional feedbacks between 5-HT and insulin imbalances that are associated with energy and emotion [6–9]. It attempts to explain some of the

1 Insulin pools throughout the body, excluding pools in the central nervous system.

2 Serotonin pools in the central nervous system, separate from serotonin pools in the gut.

underlying limitations to psychological resilience commonly experienced by persons with compromised insulin function, and to explain potential mechanistic links behind highly comorbid depression and metabolic disease.

1.1 Comorbid depression and metabolic disease

Diabetes mellitus is a metabolic condition characterized by chronic high blood sugar levels due to an underlying impairment in insulin secretion (Type 1, T1D), action (Type 2, T2D) or both [10]. Metabolic syndrome – the co-occurrence of insulin resistance, obesity, atherogenic dyslipidemia, and hypertension – is a high-risk factor for the development of T2D [11]. Diabetes is commonly comorbid with mood disorders linked to altered central 5-HT, and most often with disorders involving depression [12–14]. Diabetics are two times more likely to have depression compared to non-diabetics [15–17]. This comorbid relationship between mental health disorders and both types of diabetes appears to be bidirectional, and their co-occurrence tends to exacerbate symptom severity [12, 18, 19].

Depression refers to any of several depressive or mood disorders outlined in the DSM[3] handbook, broadly involving extended periods of sadness, emptiness and depressed mood, and often accompanied by disturbances in cognition, sleep and energy levels [18, 20, 21]. There is less clarity and consensus on underlying impairments involved in depression. Since the 1960s, depression has most commonly been explained by a deficiency in brain 5-HT activity [22, 23]. This paralleled industry's need to allocate pharmaceutical prescriptions according to symptomology [24] and thus their marketing of selective serotonergic medications [25, 26]; most commonly, 5-HT reuptake inhibitors (SSRIs) which function to increase synaptic 5-HT and upregulate 5-HT neurotransmission [7].

The neurobiological basis and treatment of depression is still the predominant narrative accepted by the Western public and endorsed by leading research and educational materials [27, 28]. However, the idea that depression is caused strictly by low 5-HT activity is mostly rejected by experts [25, 28, 29]. Some even argue that high 5-HT activity is behind depressive states, and that this is a functional response [7]. What remains largely undisputed is the extensive evidence that changes in brain 5-HT has some role in depressive states and energy regulation.

1.2 Mechanisms behind diagnostic labels

Medical research on diabetes and depression now consider the environmental and sociocultural contexts in which these neurophysiological systems interact, and how these contexts shape our organization of disease and disorder [26, 29–31]. Sociocultural factors influence how these states are generated, whether their symptoms and syndromes are interpreted as adaptive or maladaptive, and at what point these symptoms become categorized as diseases, syndromes or disorders [24, 30–33]. Psychological disorders are subjectively categorized by non-universal, Western constructs of self, normality and value [34]. Moreover, our scientific understandings of the neurological basis for cognition, emotion and behavior have been predominantly measured on Western populations, despite the evidence that changes in neurobiology lead to different expressions in different cultures [35].

---

[3] Diagnostic and Statistical Manual of Mental Disorders by the American Psychological Association.

The RDoC system from the National Institute of Mental Health is a more recent and systemic tool for categorizing mental health by functional constructs that represent a specified functional dimension of behavior [36]. The constructs are systems of response to the environment (e.g., positive valence, or reward system) which are then characterized in aggregate by the genes, molecules, circuits and behaviors involved. The RDoC thus examines the mechanism that "drive psychiatric symptoms" [37].

1.3 Peripheral insulin impairment and central serotonin availability

In light of these evolving explanations, this chapter focuses on mechanistic links between insulin and 5-HT systems that *drive* common *symptoms* of metabolic and mood dysfunction, which may shed light on the consequences of abnormal insulin synthesis and sensitivity, regardless of diagnostic labels. In this way, these feedbacks may be interpreted and applied to evolving understandings of disease and disorder in different biological and sociocultural contexts.

We thus use the unconventional term, *insulin impairment*, to refer to impaired insulin secretion or insulin sensitivity in any condition, and focus on its relationship to *5-HT availability* from Trp metabolism via the *5-hydroxyindole pathway*; despite the absence of this terminology and this scope of focus in neurophysiological medical research.

Experimental studies over several decades have investigated the mechanisms of Trp metabolism and the impacts of induced-diabetes on the distribution of 5-HT in distinct central and peripheral systems. Yet this literature remains largely fragmented, each representing isolated pieces of complex relationships involved in insulin-5-HT imbalance [38–42]. Martin and colleagues [43] addressed this gap in a novel review examining the links between central insulin impairment and central serotonergic activity. They highlight several mechanisms potentially linking these systems, including increased oxidative stress, inflammation, and hypothalamic-pituitary–adrenal axis activation. However, the bulk of this research emphasizes the downstream processes of insulin signaling in the brain. Martin and colleagues [43] show that insulin and 5-HT coregulate processes within the brain, including neurogenesis. Thus, the impairment of *central* insulin function in diabetic states disrupts the effectiveness of 5-HT *activity* in these pathways, and throughout the brain.

In contrast, the present chapter examines the role of *peripheral* insulin (that outside of the brain) in regulating *availability* of central 5-HT and its precursor Trp. We specifically focus on how peripheral insulin impairment reduces peripheral pools of Trp, the rate of Trp transport into the brain, and the rate of central 5-HT synthesis from available Trp. It is important to emphasize that the scope of this chapter focuses on the ultimate outcome of central 5-HT synthesis and thus the amount of 5-HT that could be available for neurotransmission, while Martin and colleagues [43] examine 5-HT neurotransmission (activity in the synapse). Similar to Martin and colleagues' [43] presentation of altered 5-HT activity, we see a multitude of pathways where the loss of normal insulin function can impair metabolic, mood and energy regulation through positive feedback.

In summary, this chapter aims to highlight the significant role of normal peripheral insulin systems in supporting 5-HT availability to the brain, and the multitude of pathways that impair normal 5-HT availability to the brain when this system is not functioning well. This can occur at varying degrees among individuals across these metabolic diseases and syndromes, and in non-diseased states.

2. Tryptophan metabolism

Tryptophan (Trp) is an essential amino acid that serves as a building block for 5-HT (5-HT), kynurenic acid, melatonin, and quinolinic acid, among other compounds. Imbalances in Trp levels have been associated with various diseases and disorders, including cancer, dementia, diabetes, and depression [44, 45]. Most notably, individuals with diabetes, depression, or both conditions tend to have lower levels of Trp in their blood [46, 47]. Trp is primarily metabolized through three pathways: the indole pathway, the 5-hydroxyindole pathway (which is involved in 5-HT production), and the kynurenine pathway (which converts 90% of Trp into other compounds) [2, 48].

2.1 Central tryptophan regulation

The central and peripheral pools of 5-HT are separated by the blood-brain barrier (BBB): a protective layer between the brain and blood vessels. Unlike 5-HT, Trp can cross the BBB (**Figure 1**) through a specific transporter called the large neutral amino acid transporter (LAT1) [49]. However, other large neutral amino acids (LNAA), such as tyrosine, phenylalanine, leucine, isoleucine, and valine, also compete for entry into the brain using the same transporter [50–52]. When there are high levels of these competing LNAA in the blood, the ratio of Trp to LNAA decreases, leading to reduced availability of Trp in the central nervous system [50–54]. Since Trp and other LNAAs[4] are essential amino acids, the composition of our diet strongly influences the ratio of Trp to LNAA in the blood and subsequently affects Trp uptake in the brain [53, 55, 56]. Studies have shown that elevated levels of Trp in the blood after a meal can increase peripheral Trp:LNAA [57], which is followed by elevated levels of Trp and 5-HT in the central nervous system [58, 59].

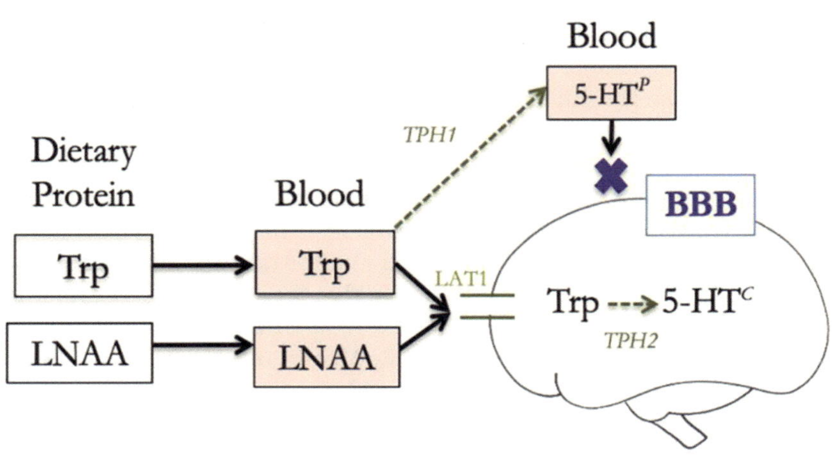

Figure 1.
Tryptophan transport past the blood brain barrier determines tryptophan availability for central serotonin synthesis (5-HT^C = central serotonin, 5-HT^P = peripheral serotonin, Trp = tryptophan, LNAA = large neutral amino acid, BBB = blood brain barrier, TPH = tryptophan hydroxylase).

[4] Except tyrosine.

2.2 Peripheral tryptophan regulation

The ratio of Trp to other large neutral amino acids (LNAA) in our diet is particularly important because Trp is the most limited LNAA, usually comprising only 1–2% of total protein [60–62].

When high-Trp proteins are consumed alone, they significantly increase the levels of Trp in the blood and the Trp to LNAA ratio compared to low-Trp proteins [39, 63, 64]. However, due to the higher concentrations of other LNAA in most dietary proteins, a high-carbohydrate meal actually increases the Trp to LNAA ratio more than a high-protein meal [65, 66]. This unexpected effect is due to the role of insulin in modulating Trp in the peripheral blood, and the stronger insulinogenic capacity of carbohydrates.

Insulin plays a crucial role in facilitating Trp uptake into the brain (**Figure 2**). Insulin helps transport LNAA into skeletal tissues, which reduces competing LNAAs in the blood, increases Trp:LNAA, and reduces competition for Trp to enter the brain [40, 67]. Insulin also helps transport Trp into skeletal tissues, but a larger proportion of branched-chain amino acids (BCAA) including valine, isoleucine, and leucine are taken up by muscle tissues in response to insulin. As a result, there is an elevated Trp:LNAA in the blood that is proportional to the circulating insulin levels [57, 68].

Insulin does impact central Trp uptake through another mechanism, but this process does not have a significant effect on Trp availability in the central nervous system. To cross the blood-brain barrier via the LAT1 transporter, Trp must be unbound [54]. Within the limited pool of Trp in the blood, most of it is reversibly bound to plasma albumin protein, with only about 10% of Trp being unbound. Insulin comes into play in this pathway by removing non-esterified fatty acids (NEFA) from albumin, creating space for Trp to bind to albumin and reducing the amount of free Trp available for transport [69]. However, the binding of Trp to albumin is transient, so the impact on Trp availability in the central nervous system is negligible [70]. As a result of this constant state of dissociation and binding with albumin, approximately 70–80% of Trp in the blood is available to cross the blood-brain barrier [70].

2.3 Insulin impairment and tryptophan

Research below demonstrates that impaired peripheral insulin function will alter Trp homeostasis and metabolic pathways in two ways: i) by redirecting Trp to regulate immune response in hyperglycemic conditions, reducing peripheral Trp

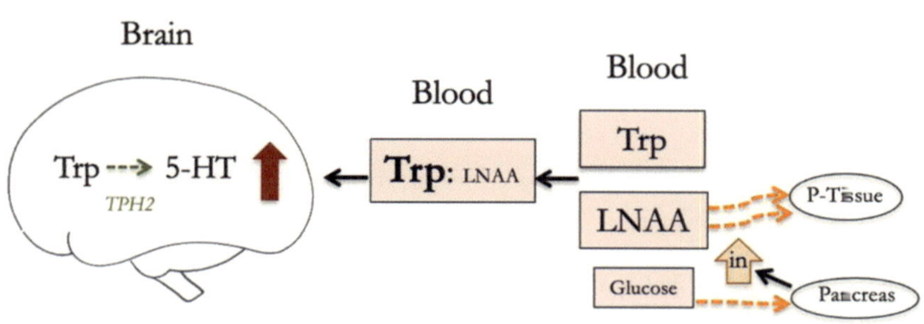

Figure 2.
Insulin facilitates tryptophan transport by increasing Trp:LNAA in the blood (5-HT = serotonin, Trp = tryptophan, LNAA = large neutral amino acid, BBB = blood brain barrier, TPH = tryptophan hydroxylase, P-tissue = skeletal tissues, in = insulin).

availability and ii) by modifying insulin's role in regulating amino acid ratios, limiting Trp transport into the brain. Multiple pathways leading to reduced Trp levels and altered Trp metabolism in diabetic states have been extensively studied in rodents and humans; insulin plays a crucial role in modulating these changes.

Chemical impairment of insulin production in rodents[5] resulted in a significant decrease in total plasma Trp after 7 day [42], 28 days [40], and 35 days [71]. Similar findings have been reported in human studies. Three studies involving children with clinical type 1 diabetes (T1D) found lower plasma Trp and Trp:LNAA levels compared to non-diabetics [38, 72, 73]. Adolescents with metabolic syndrome also showed lower peripheral Trp levels compared to controls [74, 75]. In a study involving healthy controls and diabetic adults individuals with T1D had significantly lower serum Trp levels compared to the control group [76]. Chemically induced acute T1D in rodents also led to a significant increase in plasma branched-chain amino acids (BCAAs) valine, leucine, and isoleucine, resulting in a decrease in the overall Trp:LNAA [71, 77] and subsequently reduced central Trp levels [40, 71, 77–79]. Human studies have also found low Trp:LNAA in individuals with diabetes [77, 80].

The culmination of this research shows that impaired insulin function and insulin deficiency limit the uptake of Trp competitors into skeletal tissues, maintain low Trp:LNAA levels and LNAA competition for LAT1, and thus result in lower central Trp levels (**Figure 3**). Importantly, elevated peripheral BCAA concentrations have been associated with an increased risk of future diabetes and may indicate impaired insulin function even before the diagnosis of metabolic disorders [81, 82].

Studies in rodents and humans have shown that while initial insulin impairment increases unbound peripheral trp chronic stages of insulin impairment eventually led to a significant decrease in plasma Trp, further reducing the Trp:LNAA and central Trp levels [83, 84]. Chronic insulin impairment can affect peripheral Trp levels through changes in Trp metabolism and oxidation. Impaired insulin function disrupts normal glucose uptake, leading to hyperglycemic states that cause inflammation and

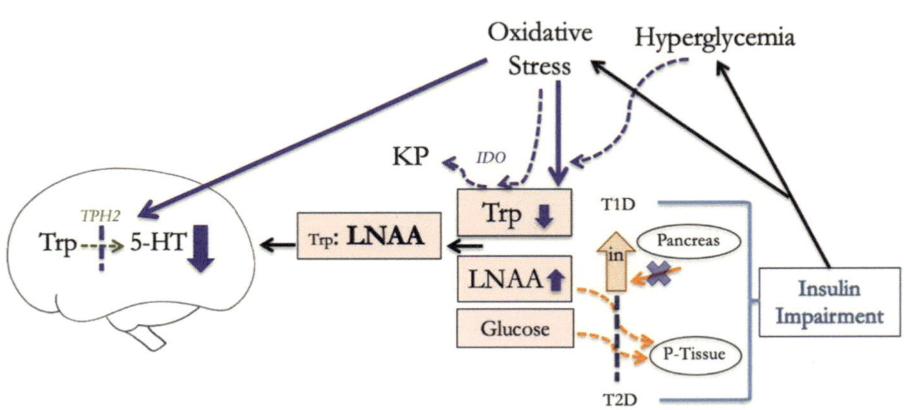

Figure 3.
Insulin impairment lowers serotonin production in the brain by impaired ability to elevate peripheral Trp:LNAA and facilitate central Trp uptake, and by alteraing TPH2 enzyme activity (5-HT = serotonin, Trp = tryptophan, LNAA = large neutral amino acid, TPH2 = tryptophan hydroxylase 2, P-tissue = skeletal tissues, in = insulin, KP = kyneurenine pathway, T1D = type 1 diabetes, T2D = type 2 diabetes, IDO = indoleamine 2,3-dioxygenase).

[5] Streptozotocin (STZ) impairs normal insulin production by damaging pancreatic cells.

increased production of reactive oxygen species by mitochondria [85–87]. Chronic inflammation and oxidative stress trigger an immune response that upregulates Trp metabolism in the kynurenine pathway, ultimately reducing peripheral Trp pools (**Figure 3**) [72, 88–91]. This also has direct impacts on Trp levels because the highly reactive indole ring of Trp can be easily damaged by oxidizing species [73, 92].

New research suggests a connection between plasma Trp and peripheral insulin through the GPR142 receptor located on pancreatic islet cells. GPR142 is a G-protein coupled receptor that specifically binds to Trp and phenylalanine [77]. Binding of Trp to GPR142 stimulates the release of insulin, glucagon-like peptide-1 (GLP-1), and glucose-dependent insulinotropic polypeptide (GIP), helping to regulate glucose levels [93]. Administration of Trp in obese mice significantly increased glucose metabolism and insulin secretion [94]. In GPR142 knockout mice, Trp supplementation did not lead to increased insulin release or improved glucose tolerance compared to controls, highlighting the importance of GPR142 in the Trp-insulin link [94]. The action of Trp on GPR142 receptors in the pancreas creates a positive feedback loop where chronically reduced Trp levels in impaired insulin states may further decrease insulin secretion, disrupt glucose homeostasis, and subsequently lower plasma Trp levels.

3. Serotonin synthesis

5-HT is synthesized from tryptophan (Trp) through the 5-hydroxyindole metabolic pathway (**Figure 3**). This process is catalyzed by either Trp hydroxylase enzyme 1 (TPH1) or Trp hydroxylase enzyme 2 (TPH2). TPH1 is primarily found in peripheral tissues, while TPH2 is responsible for converting Trp to 5-HT in the brain. The rate of 5-HT synthesis is dependent on the activity of the TPH enzyme. In the brain, the synthesis of the intermediate molecule 5-hydroxytryptophan (5-HTP) is the limiting step, and this step is determined by the availability of Trp as a substrate.

3.1 Insulin impairment and serotonin

Insulin impairment not only affects central 5-HT synthesis by altering enzyme activity in the 5-hydroxyindole pathway, but it also impacts the catalytic function of TPH2. In rodents with chemically-induced insulin impairment for 7 days, TPH2 activity was reduced in the cerebral cortex and brainstem [89]. This decrease in activity was previously attributed solely to low central Trp availability as a result of the mechanisms described earlier. However, this same study found that the TPH2 enzyme in diabetic rodents had a significantly lower affinity for Trp and reduced phosphorylating capacity required to stimulate the enzyme. Another study with T1D rodents observed similar catalytic dysfunction of TPH2, as well as reduced expression of the enzyme [42]. These epigenetic and kinetic changes were attributed to hyperglycemic conditions: elevated peripheral glucose levels (hyperglycemia) led to increased brain glucose levels, which can trigger inflammation, reactive oxygen species, and oxidative stress, ultimately causing damage to TPH2 expression and function [95–98].

Chronic elevation of glucocorticoid levels, commonly seen in diseases associated with insulin impairment, can also explain the reduced expression of TPH2 [99, 100]. Excess glucocorticoids are released in response to chronic stress through activation of the hypothalamic–pituitary–adrenal axis. Glucocorticoid receptors act as transcription factors that inhibit TPH2 expression and can significantly decrease 5-HT levels in the raphe nuclei after one week of excess exposure [101]. Elevated glucocorticoid levels are also independently linked to peripheral tissue insulin resistance, impaired insulin production, and high blood glucose levels [99, 100, 102]. Overall, chronic release of glucocorticoids appears to indirectly impair central 5-HT availability by altering Trp transport and conversion, as well as via hyperglycemic conditions. The correlation between elevated glucocorticoids and depressive symptoms highlights this stress response as a contributing factor in the disrupted regulation of Trp and 5-HT, which underlies psycho-metabolic comorbidities [103]. In summary, both reduced central Trp availability and impaired TPH2 activity disrupt the 5-hydroxyindole pathway in individuals with chronic insulin impairment.

The effects of altered Trp metabolism in response to peripheral insulin impairment, such as lower peripheral Trp, elevated peripheral LNAA, and lower central TPH2 expression and function, are expected to subsequently reduce central 5-HT pools available for neurotransmission. This prediction is supported by several rodent studies, where chemically induced insulin impairment led to a significant decrease in central 5-HT synthesis. This decrease was observed after 4 weeks in the whole brain [78], after 1 week in the whole brain [104], after 2 weeks in the whole brain [41] and the hypothalamus [89], and after 50 days in the striatum and pons medulla [105], as indicated by 5-HTP accumulation. Additionally, 5-HT synthesis measured by TPH2 activity was significantly lower after 7 days in the cerebral cortex and brainstem [89]. While this is beyond the scope of the present chapters, readers may consider that insulin impairment is also expected to compromise other systems dependent on Trp and 5-hydroxyindole metabolism of Trp. For example, the synthesis of the 5-HT metabolite, melatonin (MLT), appears to be significantly reduced in diabetic states [6].

3.2 Insulin treatment

Insulin treatments in diabetic rodents have a limited capacity to restore normal Trp and 5-HT levels in the brain. Prolonged periods of hyperglycemia place greater stress on both insulin secretion and insulin sensitivity, further reducing the potential for recovery [106]. A comprehensive study examined the effects of insulin treatment on rodents with 7 days of insulin impairment by measuring TPH2 activity, central 5-HT levels, central Trp levels, and blood glucose levels [42]. Following insulin treatment, rodents with induced T1D showed normal blood glucose and Trp levels in the brain, but TPH2 activity and 5-HT levels remained depressed after 7 and 14 days of insulin treatment [42]. However, These findings are contradicted by a different study of rodents with 7 days of insulin impairment, where insulin treatment after 14 days was able to restore near-normal central 5-HT levels [104].

A study from 1991 suggested that prolonged insulin impairment may result in irreversible reduction in central 5-HT levels [77]. Rodents with 10–30 days of insulin impairment were treated with insulin at various time intervals ranging from 15 to 135 days. Insulin treatment restored serum Trp levels throughout all time spans but could not restore central Trp levels beyond 15 days. It is important to note that this study induced insulin impairment using alloxan, which is known to cause higher cellular toxicity in non-target cells compared to streptozotocin used in other studies [42, 77, 104, 107].

Human studies also offer insights into the consequences long term insulin impairment, as in the case of clinically diagnosed diabetes. In comparison to normal controls, diabetics had higher levels of plasma BCAA [108], tyrosine and phenylalanine [109], and lower levels of plasma Trp [110]. The subsequently low plasma Trp:LNAA was not reversed by weight loss [111]. Diversion of Trp metabolism to the kynurenine pathway in response to stress persisted after bariatric surgery [112] and plasma Trp and Trp:LNAA were little changed by Trp dosing in obese persons compared to lean [109].

Both rodent and human studies indicate that chronic peripheral insulin impairment may have long-term effects on 5-HT availability that cannot be fully restored by insulin treatment, underscoring the complexity of identifying effective treatments.

4. Behavioral and psychotropic treatments

The dysregulation of insulin and serotonergic systems is further complicated by a positive feedback loop involving common behavioral responses and psychotropic medical treatments. For instance, central 5-HT depletion can lead to cravings for carbohydrates in an attempt to raise peripheral insulin levels [49, 113–115]. However, a high-glycemic diet increases the risk of insulin resistance, impaired insulin secretion, and depression [106, 116–118]. Carbohydrate cravings are also common in various disorders associated with mood changes and serotonergic modulation, such as atypical depression, seasonal affective disorder, late luteal phase dysphoric disorder, and binge eating disorder. Serotonergic medications that upregulate 5-HT neurotransmission are often prescribed to treat these disorders [119, 120].

However, these medications may exacerbate Trp and 5-HT imbalances in individuals with peripheral insulin impairment. As serotonergic medications are not specific to the central nervous system, they disrupt 5-hydroxyindole homeostasis in the periphery, where most 5-HT is synthesized [121]. This can lead to apoptosis of pancreatic cells and acute pancreatitis (inflammation and tissue damage), and thereby inhibit insulin secretion [122, 123]. It can also impair insulin receptor function [124]. Several studies have observed an increased risk of type 2 diabetes associated with SSRI use [125–128], despite the effect of SSRIs on glucose homeostasis [118], and this risk seems to dose, concentration and duration dependent [127, 128]. Moreover, individuals with pre-existing low levels of peripheral Trp are more susceptible to anxious and depressive episodes when taking serotonergic antidepressants (**Figure 4**) [129, 130].

In summary, the regulation of Trp and 5-HT homeostasis is crucial in both the periphery and brain. Insulin impairment and common attempts to regulate emotional changes

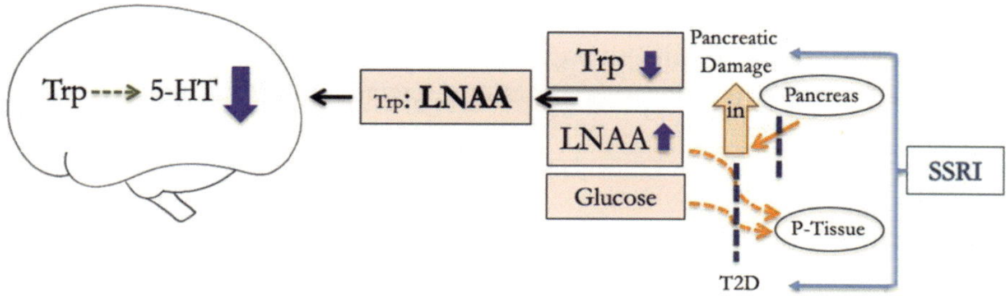

Figure 4.
Selective serotonin reuptake inhibitors (SSRI) disrupt homeostatic regulation of serotonin in the periphery, which can restrict peripheral insulin secretion and sensitivity, reducing available central 5-HT (5-HT = serotonin, Trp = tryptophan, LNAA = large neutral amino acid, P-tissue = skeletal tissues, in = insulin, KP = kyneurenine pathway, T1D = type 1 diabetes, T2D = type 2 diabetes).

caused by this impairment can disrupt this delicate balance. The evidence from rodent and human studies presented in this chapter suggests that behavioral and medical treatments that exacerbate insulin impairment also pose a risk of long-term and potentially irreversible impairment of Trp metabolism and transport, as well as central 5-HT availability.

5. Interpretations and applications

How do we interpret the consequences of low Trp levels and restricted central 5-HT in light of positive feedbacks with higher degree and duration of insulin impairment? Returning to ideas presented in the Introduction, what do these interactions and outcomes mean within a new paradigm that dissociates neurobiology from absolute diagnostic labels and questions the 5-HT 'chemical imbalance'? Low and high 5-HT hypotheses, and the medications targeting neurotransmission upregulation, are not modeled on balance. Rather, they are modeled on extremes and deficits of a single chemical and blind targeting of multiple systems in the body [131], all in an attempt to chronically maintain ideal emotional states.

5.1 Complex adaptive living systems

These mechanisms can instead be viewed with a model of balance that better reflects the reality of functional living systems. Optimization of single brain chemicals is reflective of a linear cause-effect reductionist approach to health [132, 133]. Alternatively, we interpret the patterns of interaction among neurological and metabolic systems with the view that human beings are composed of, and operate within, complex adaptive systems (CAS) [132–134].

A complex adaptive system is a collection of specialized agents (components or parts) that can be understood within the context of the whole and the interconnecting systems they comprise [133, 134]. Diverse interactions between these systems and environmental stimuli will adapt to give rise to non-linear, unpredictable and ever-changing dynamics [26, 135, 136]. While defined by dynamic change and evolution, complex living systems conserve their conditions for renewability through self-regulating feedback loops that link interacting agents [134, 137]. Collectively, these traits allow CAS to maintain homeostatic equilibrium such that they maintain a dynamic form of stability over time [138] and develop resiliency in the face of disturbance and adversity [133–135, 137].

This CAS perspective is especially useful in understanding complex systems that are not easily measured, understood and predicted [132]. Applying the CAS perspective to the human brain and body, we view health as homeostatic equilibrium of essential interactions, systems and functions needed for sustaining the human as a self-organizing system. This lens can be applied to neurophysiological systems by viewing them as normally fluctuating between order and disorder, and by focusing on "state change" rather than static maintenance of absolute highs and lows [132]. We interpret positive feedback loops with their "tend[ency] towards chaos and decay" and identify a need to counterbalance by negative feedback [132]. The CAS model thus shifts our view of neurophysiological imbalance from the failure of a single component in a particular place, to the idea that imbalance arises from interaction between multiple systems that are not able to self-regulate and maintain dynamic equilibrium [132, 133]. It also recognizes that these interactions will generate emergent, unpredictable outcomes that manifest differently in unique individuals and sociocultural contexts [133, 137]. We apply these principles to our understanding of the insulin system as it interacts with 5-hydroxyindole system in the body and brain.

5.2 Systems of Insulin and 5-hydroxyindole metabolism

Functional homeostatic regulation occurs when components of the system are maintained within upper and lower bounds, and fluctuate dynamically in response to changes in their environment [24, 26, 135, 139]. Indeed, both an excess and deficiency of 5-HT and its precursor throughout the body can lead to health imbalances [2]: sufficiently low peripheral 5-HT is needed to maintain insulin sensitivity and prevent obesity [140, 141], yet sufficiently high levels of peripheral 5-HT are needed to maintain pancreatic insulin secretion [142] and normal glucose levels [143]. Similarly, a "high" Trp diet in human studies has been labeled both a protective [144] and risk factor [83] for later development of T2D.

A state of acute tryptophan depletion (ATD) can be experimentally induced to explore the behavioral and cognitive responses to low central 5-HT. Reduced central 5-HT from ATD led not only to negative affect bias (sad mood), but also better punishment prediction accuracy and more risk aversive behavior [145–147]; it also enhanced negative reciprocity, expressed as greater punishment or retaliation in response to perceived unfairness [148]. While these findings tend to be interpreted as explanations of maladaptive depressive states, we can revisit this from a CAS perspective. Low central 5-HT states may be an adaptive, functional response to avoid harm that is learned in environments where punishment is common and then recruited when complex social issues are interpreted as chronic danger. Other scholars predict that hypervigilance is an adaptive response for a child raised in an unpredictable environment [32], and that social avoidance is an adaptive response to volatile social experiences [149].

Evolutionary psychologists tend to view short periods of stress, anxiety or depression as adaptive responses that can increase resiliency in certain contexts if the triggers are addressed [7, 147, 149, 150]. For example, the *Analytical Rumination Hypothesis* posits that depressive symptoms of cognitive rumination and loss of interest (lack of concentration, appetite, sex drive or socialization) are coordinated as an evolutionary adaptation to solve complex social problems or traumatic experiences with sustained focus [7, 151, 152]. These researchers show overlapping genetic, neurobiological and symptom expression between depressive symptoms and sickness behavior, with the former acting as an emotional fever. Just as our bodies divert energy towards immune function in response to a viral threat at the expense of growth and reproduction, biological trade-offs occur during depression, with a reallocation of resources in response to persistent environmental threats.

Both extremes of lowered central 5-HT availability and heightened 5-HT neurotransmission impair the brain's ability to accurately interpret facial emotions; ATD inaccurately escalated the interpretation of fearful emotion to anger, while SSRI's dissolved the ability to distinguish fearful and normal emotions [145]. These results support the idea that functional 5-hydroxyindole metabolism and 5-HT neurotransmission may require dynamic homeostatic regulation.

From a resiliency perspective, ATD alone does not seem to initiate the cascade of positive feedbacks in insulin and 5-hydroxyindole systems; it worsens or triggers them. In fact, Trp-deficiency induced episodes of anxiety and depression in people with a personal or familial history of mental health disorder, yet had little to no emotional effect on healthy controls [153–155]. Similarly, in comparisons of high-Trp versus low-Trp consumption (where high-Trp levels significantly elevated plasma Trp and Trp:LNAA), high Trp improved memory exclusively in individuals susceptible to high stress [64], and reduced vulnerability to experimental triggers for fatigue, negative affective bias, and diminished well-being [63].

It is important to recognize that these ATD tests represent acutes stress, while stress responses outside of the laboratory are often triggered by more complex, abstract threats and exist in a more pervasive form that requires reflection on previous experiences and learned behavior [146, 147]. The chronic nature of these real stressors can cause damage to the brain and body that correspond with mental unwellness and symptoms of non-communicable diseases [156, 157].

Interpreting these experiments through CAS perspective, we suggest that recovery from fluctuations in mood and depressive episodes may be supported by the adaptive capacity to fluctuate high and low neurotransmission in response to biosocial environmental disturbances. 5-HT availability is a limiting factor in the potential to upregulate neurotransmission at a given time and in response to a given situation. Availability also determines the potential to downregulate in a time-responsive manner relative to upregulation, as needed to re-establish equilibrium.

5.3 Serotonin as resilience molecule

This idea is supported by evolutionary studies of 5-hydroxyindole metabolism arguing that a tightly regulated balance of Trp content and availability is most ideal for all living organisms in order to maintain the adaptive function of its intermediates [158]. The human body has developed several mechanisms for maintaining a narrow range of Trp in different regions [158] and in relation to co-existing amino acids [2]. The 5-hydroxyindole pathway that produces 5-HT from Trp has been highly conserved from unicellular bacteria to mammals, and shares one common function across all organisms and tissues: adaptive response to environmental stress [158–160]. Despite the body having very low Trp concentrations compared to other amino acids, its metabolite 5-HT serves many crucial biological functions. 5-HT is considered the homeostatic regulator of the central nervous, neuroendocrine, gut and immune systems, and the biochemical connector of mind and body with the environment [8, 158].

The adaptive function of 5-HT as a moderator of stress response can be understood as a trait supporting psychological resilience: the ability to cope with or recover from adversity [161]. We argue that altered Trp metabolism as a result of chronic insulin impairment ultimately impairs psychological resiliency, particularly among already vulnerable individuals. By limiting availability of 5-HT in the brain, chronic insulin impairment disrupts the ability to finely regulate changes in 5-HT neurotransmission in response to changing environmental contexts. Since 5-HT is responsible for adaptive response to stress, this represents a loss of neurological options that will further increase vulnerability to emotional and energy disturbances.

Insulin impairment is one mechanism that impairs the adaptive capacity of Trp metabolites like 5-HT to mitigate impacts of environmental stress. The dynamic state of insulin impairment and associated glucose intolerance throughout the life course of prediabetic and diabetic individuals [10, 162, 163], as well as the diverse experiences of individuals themselves, must be considered in light of experimental studies. Moreover, our understanding of the complex and dynamic state of insulin impairment, diabetes and metabolic disorder, cannot be based on short-term rodent studies alone. In combination with human diabetic studies, these experiments do identify clear limits in physiological function during these complex states; most notably, during early states of insulin impairment that occurs long before changes to blood-glucose, weight, or energy levels are detected at diagnosis of metabolic disorder. Together, these studies demonstrate a strain in adaptive response to stress that will be affected by the severity and duration of insulin impairment and individual human characteristics.

6. Conclusion

Insulin impairment disrupts the homeostatic adaptive capacity to regulate central 5-HT and impairs psychological resilience to stress by altering normal 5-hydroxyindole metabolism of Trp. The small fraction of 5-HT in the brain is more vulnerable to insulin impairment than peripheral 5-HT since, in addition to being limited by low peripheral Trp stores, it is also restricted by impaired TPH2 activity and impaired central Trp uptake. Insulin impairment represents a loss of options for the many roles of central 5-HT, which are increasingly restricted by higher degree and duration of insulin impairment, as well as serotonergic medications and dietary cravings induced by the dysfunction itself.

Neurophysiological studies of high and low 5-HT might be better understood from the perspective homeostatic balance of 5-hydroxyindole metabolism, and how this is shaped by dynamic states of other molecules in the body, beyond insulin impairment. Future studies may elucidate our understanding of a bidirectional relationship between insulin and 5-HT function and explore the 5-hydroxyindole response to glycemic imbalance and oxidative stress under diverse conditions and locations in the brain and body (i.e., how different receptors in different areas of the brain respond to insulin impairment over time). Better integration of experimental studies that focus on the origins of neurophysiological imbalance may help identify treatment that supports adaptive capacity inherent to the 5-HT system. Most significantly, we encourage neurophysiological research to consider experimental design and interpretation with the resiliency model of complex adaptive living systems.

Acknowledgements

Thank you to the Trinity College, University of Toronto Independent Study Course for supporting this research.

Author details

Nicole Spiegelaar[1]* and Sebastian Warma[2]

1 School of the Environment and Trinity College, University of Toronto, Canada

2 Department of Health Sciences and Technology, ETH, Zürich, Switzerland

*Address all correspondence to: nicole.spiegelaar@utoronto.ca

References

[1] Azmitia E. Chapter 1.1 - evolution of serotonin: Sunlight to suicide. Handbook of Behavioral Neuroscience. 2010;**21**:3-22

[2] Sainio EL, Pulkki K, Young SN. L-tryptophan: Biochemical, nutritional and pharmacological aspects. Amino Acids. 1996;**10**(1):21-47

[3] Berger M, Gray JA, Roth BL. The expanded biology of serotonin. Annual Review of Medicine. 2009;**60**:355-366

[4] Kraus C, Castrén E, Kasper S, Lanzenberger R. Serotonin and neuroplasticity - links between molecular, functional and structural pathophysiology in depression. Neuroscience and Biobehavioral Reviews. 2017;**77**:317-326

[5] Yabut JM, Crane JD, Green AE, Keating DJ, Khan WI, Steinberg GR. Emerging roles for serotonin in regulating metabolism: New implications for an ancient molecule. Endocrine Reviews. 2019;**40**(4):1092-1107

[6] Amaral FG, Turati AO, Barone M, Scialfa JH, Buonfiglio D, Peres R, et al. Melatonin synthesis impairment as a new deleterious outcome of diabetes-derived hyperglycemia. Journal of Pineal Research. 2014;**57**(1):67-79

[7] Andrews PW, Bharwani A, Lee KR, Fox M, Thomson JA. Is serotonin an upper or a downer? The evolution of the serotonergic system and its role in depression and the antidepressant response. Neuroscience and Biobehavioral Reviews. 2015;**51**:164-188

[8] Azmitia EC. Modern views on an ancient chemical: Serotonin effects on cell proliferation, maturation, and apoptosis. Brain Research Bulletin. 2001;**56**(5):413-424

[9] Heisler LK, Kanarek RB, Gerstein A. Fluoxetine decreases fat and protein intakes but not carbohydrate intake in male rats. Pharmacology, Biochemistry, and Behavior. 1997;**58**(3):767-773

[10] ADA (American Diabetes Association). Diagnosis and classification of diabetes mellitus. Diabetes Care. 2014;**36**(1):S67-S74

[11] Huang PL. A comprehensive definition for metabolic syndrome. Disease Models & Mechanisms. 2009;**2**(5-6):231-237

[12] Balhara YPS. Diabetes and psychiatric disorders. Indian Journal of Endocrinology Metabolism. 2011;**15**(4):274-283

[13] Toalson P, Ahmed S, Hardy T, Kabinoff G. The metabolic syndrome in patients with severe mental illnesses. Primary Care Companion Journal of Clinical Psychiatry. 2004;**6**(4):152-158

[14] Canadian Diabetes Association Clinical Practice Guidelines Expert Committee, Robinson DJ, Luthra M, Vallis M. Diabetes and mental health. Can Journal of Diabetes. 2013;**37**(Suppl. 1):S87-S92

[15] Fisher L, Skaff MM, Mullan JT, Arean P, Glasgow R, Masharani U. A longitudinal study of affective and anxiety disorders, depressive affect and diabetes distress in adults with type 2 diabetes. Diabetic Medicine. 2008;**25**(9):1096-1101

[16] Egede LE, Zheng D, Simpson K. Comorbid depression is associated with increased health care use and expenditures in individuals with diabetes. Diabetes Care. 2002;**25**(3):464-470

[17] Hermanns N, Kulzer B, Krichbaum M, Kubiak T, Haak T. Affective and anxiety disorders in a German sample of diabetic patients: Prevalence, comorbidity and risk factors. Diabetic Medicine. 2005;**22**(3):293-300

[18] Pan A, Lucas M, Sun Q, van Dam RM, Franco OH, Manson JE, et al. Bidirectional association between depression and type 2 diabetes mellitus in women. Archives of Internal Medicine. 2010;**170**(21):1884-1891

[19] Musselman DL, Betan E, Larsen H, Phillips LS. Relationship of depression to diabetes types 1 and 2: Epidemiology, biology, and treatment. Biological Psychiatry. 2003;**54**(3):317-329

[20] Chand SP, Arif H. Depression. Treasure Island, FL: StatPearls Publishing; 2023. Available from: http://www.ncbi.nlm.nih.gov/books/NBK430847/

[21] Rose AL, Hopko DR, Lejuez CW, Magidson JF. Major depressive disorder. In: Functional Analysis in Clinical Treatment. 2nd ed. San Diego, CA, US: Elsevier Academic Press; 2020. pp. 339-373

[22] Albert PR, Benkelfat C, Descarries L. The neurobiology of depression—Revisiting the serotonin hypothesis. I. Cellular and molecular mechanisms. Philosophical Transactions of the Royal Society of London. Series B, Biological Sciences. 2012;**367**(1601):2378-2381

[23] Coppen A. The biochemistry of affective disorders. The British Journal of Psychiatry. 1967;**113**(504):1237-1264

[24] Kirmayer LJ, Gomez-Carrillo A, Veissière S. Culture and depression in global mental health: An ecosocial approach to the phenomenology of psychiatric disorders. Social Science & Medicine. 2017;**183**:163-168

[25] Cowen PJ, Browning M. What has serotonin to do with depression? World Psychiatry. 2015;**14**(2):158-160

[26] Kirmayer LJ. Re-visioning psychiatry: Toward an ecology of mind in health and illness. In: Kirmayer LJ, Lemelson R, Cummings CA, editors. Re-Visioning Psychiatry. Cambridge: Cambridge University Press; 2015. pp. 622-660. Available from: https://www.cambridge.org/core/product/identifier/9781139424745%23CT-bp-24/type/book_part

[27] Pescosolido BA, Martin JK, Long JS, Medina TR, Phelan JC, Link BG. "A disease like any other"? A decade of change in public reactions to schizophrenia, depression, and alcohol dependence. AJP. 2010;**167**(11):1321-1330

[28] Moncrieff J, Cooper RE, Stockmann T, Amendola S, Hengartner MP, Horowitz MA. The serotonin theory of depression: A systematic umbrella review of the evidence. Molecular Psychiatry. 2023;**28**(8):3243-3256

[29] Pies R. Psychiatry's New Brain-Mind and the Legend of the "Chemical Imbalance." Psychiatric Times [Internet]. 2011 [cited 2023 Nov 7]. Available from: https://www.psychiatrictimes.com/view/psychiatrys-new-brain-mind-and-legend-chemical-imbalance

[30] Chatterton C. Metabolic syndrome: The construction of a "New" medical problem and the socio-ethical consequences [Internet] [PhD]. Cardiff University. 2014 [cited 2023 Nov 5]. Available from: https://orca.cardiff.ac.uk/id/eprint/58973/

[31] Simmons RK, Alberti KGMM, Gale EAM, Colagiuri S, Tuomilehto J, Qiao Q, et al. The metabolic syndrome: Useful concept or clinical tool? Report

of a WHO expert consultation. Diabetologia. 2010;**53**(4):600-605

[32] Haltigan J. Rethinking mental illness – and health. CAMH Discovers [Internet]. 2018 [cited 2020 Jun 30]. Available from: https://www.camh.ca/en/camh-news-and-stories/rethinking-mental-illness-and-health

[33] Preskorn SH, Baker B. The overlap of DSM-IV syndromes: Potential implications for the practice of Polypsychopharmacology, psychiatric drug development, and the human genome project. Journal of Psychiatric Practice®. 2002;**8**(3):170

[34] La Roche MJ, Fuentes MA, Hinton D. A cultural examination of the DSM-5: Research and clinical implications for cultural minorities. Professional Psychology: Research and Practice. 2015;**46**(3):183-189

[35] Kim HS, Sherman DK, Taylor SE, Sasaki JY, Chu TQ, Ryu C, et al. Culture, serotonin receptor polymorphism and locus of attention. Social Cognitive and Affective Neuroscience. 2010;**5**(2-3):212-218

[36] National Institute of Mental Health. Research Domain Criteria (RDoC) [Internet]. 2015 [cited 2023 Feb 10]. Available from: https://www.nimh.nih.gov/research/research-funded-by-nimh/rdoc

[37] Sanislow CA, Pine DS, Quinn KJ, Kozak MJ, Garvey MA, Heinssen RK, et al. Developing constructs for psychopathology research: Research domain criteria. Journal of Abnormal Psychology. 2010;**119**(4):631-639

[38] Herrera R, Manjarrez G, Nishimura E, Hernandez J. Serotonin-related tryptophan in children with insulin-dependent diabetes. Pediatric Neurology. 2003;**28**(1):20-23

[39] Fernstrom JD, Langham KA, Marcelino LM, Irvine ZLE, Fernstrom MH, Kaye WH. The ingestion of different dietary proteins by humans induces large changes in the plasma tryptophan ratio, a predictor of brain tryptophan uptake and serotonin synthesis. Clinical Nutrition. 2013;**32**(6):1073-1076

[40] MacKenzie RG, Trulson ME. Effects of insulin and streptozotocin-induced diabetes on brain tryptophan and serotonin metabolism in rats. Journal of Neurochemistry. 1978;**30**(1):205-211

[41] Crandall EA, Gillis MA, Fernstrom JD. Reduction in brain serotonin synthesis rate in streptozotocin-diabetic rats. Endocrinology. 1981;**109**(1):310-312

[42] Manjarrez-Gutierrez G, Neri-Gómez T, Herrera-Márquez R, Mondragón J, Oca A, Rodríguez J. Brain serotonergic disturbances caused by diabetes mellitus are not reversed by insulin treatment. 2015

[43] Martin H, Bullich S, Martinat M, Chataigner M, Di Miceli M, Simon V, et al. Insulin modulates emotional behavior through a serotonin-dependent mechanism. Molecular Psychiatry. 2022;7:1-10

[44] Platten M, Wick W, Van den Eynde BJ. Tryptophan catabolism in cancer: Beyond IDO and tryptophan depletion. Cancer Research. 2012;**72**(21):5435-5440

[45] Lovelace MD, Varney B, Sundaram G, Lennon MJ, Lim CK, Jacobs K, et al. Recent evidence for an expanded role of the kynurenine pathway of tryptophan metabolism in neurological diseases. Neuropharmacology. 2017;**112**(Pt B):373-388

[46] Manjarrez-Gutierrez G, Marquez RH, Mejenes-Alvarez SA,

Godinez-Lopez T, Hernandez-R J. Functional change of the auditory cortex related to brain serotonergic neurotransmission in type 1 diabetic adolescents with and without depression. The World Journal of Biological Psychiatry. 2009;**10**(4 Pt 3):877-883

[47] Delgado I, Cussotto S, Anesi A, Dexpert S, Aubert A, Aouizerate B, et al. Association between the indole pathway of tryptophan metabolism and subclinical depressive symptoms in obesity: A preliminary study. International Journal of Obesity. 2022;**46**(4):885-888

[48] Cheng Y, Li Y, Benkowitz P, Lamina C, Köttgen A, Sekula P. The relationship between blood metabolites of the tryptophan pathway and kidney function: A bidirectional Mendelian randomization analysis. Scientific Reports. 2020;**10**(1):12675

[49] Cansev M, Wurtman R. Aromatic amino acids in the brain. In: Handbook of Neurochemistry and Molecular Neurobiology. New York, NY: Springer Science+Business Media; 2007. pp. 59-97. Available from: https://avesis.uludag.edu.tr/yayin/81facf74-5f8e-4748-9889-8665f7736eac/handbook-of-neurochemistry-and-molecular-neurobiology

[50] Pietz J, Kreis R, Rupp A, Mayatepek E, Rating D, Boesch C, et al. Large neutral amino acids block phenylalanine transport into brain tissue in patients with phenylketonuria. The Journal of Clinical Investigation. 1999;**103**(8):1169-1178

[51] Yanagida O, Kanai Y, Chairoungdua A, Kim DK, Segawa H, Nii T, et al. Human L-type amino acid transporter 1 (LAT1): Characterization of function and expression in tumor cell lines. Biochimica et Biophysica Acta. 2001;**1514**(2):291-302

[52] Zepf F. Principles of rapid tryptophan depletion and its use in research on neuropsychiatric disorders. In: Amino Acids in Human Health and Nutrition. Oxfordshire: CABI; 2012. pp. 418-426

[53] Moja EA, Stoff DM, Gessa GL, Castoldi D, Assereto R, Tofanetti O. Decrease in plasma tryptophan after tryptophan-free amino acid mixtures in man. Life Sciences. 1988;**42**(16):1551-1556

[54] Yuwiler A, Oldendorf WH, Geller E, Braun L. Effect of albumin binding and amino acid competition on tryptophan uptake into brain. Journal of Neurochemistry. 1977;**28**(5):1015-1023

[55] Feurté S, Gerozissis K, Regnault A, Paul FM. Plasma Trp/LNAA ratio increases during chronic ingestion of an alpha-lactalbumin diet in rats. Nutritional Neuroscience. 2001;**4**(5):413-418

[56] Shen YB, Voilqué G, Odle J, Kim SW. Dietary L-tryptophan supplementation with reduced large neutral amino acids enhances feed efficiency and decreases stress hormone secretion in nursery pigs under social-mixing stress. The Journal of Nutrition. 2012;**142**(8):1540-1546

[57] Sainio EL, Närvänen S, Sainio P, Tuohimaa P. Distribution of L-tryptophan in normal and glucose - loaded mice. Amino Acids. 1995;**8**(2):209-216

[58] Fernstrom JD, Wurtman RJ. Brain serotonin content: Physiological regulation by plasma neutral amino acids. Science. 1972;**178**(4059):414-416

[59] Fernstrom JD, Wurtman RJ. Brain serotonin content: Increase following ingestion of carbohydrate diet. Science. 1971;**174**(4013):1023-1025

[60] Spiegelaar N, Martin ID, Tsuji LJS. Indigenous subarctic food Systems in

Transition: Amino acid composition (including tryptophan) in wild-harvested and processed meats. International Journal of Food Science. 2019;**27**(2019):1-14

[61] United States Department of Agriculture. USDA Nutrient Data Laboratory [Internet]. 2015 [cited Oct 29]. Available from: https://fnic.nal.usda.gov/food-composition/usda-nutrient-data-laboratory

[62] CFIA (Canadian Food Inspection Agency). Elements within the Nutrition Facts Table [Internet]. [cited 2014 Jul 3]. Available from: http://www.inspection.gc.ca/food/labelling/food-labelling-for-industry/nutrition-labelling/elements-within-the-nutrition-facts-table/eng/1389206763218/1389206811747?chap=8

[63] Gibson EL, Vargas K, Hogan E, Holmes A, Rogers PJ, Wittwer J, et al. Effects of acute treatment with a tryptophan-rich protein hydrolysate on plasma amino acids, mood and emotional functioning in older women. Psychopharmacology. 2014;**231**(24):4595-4610

[64] Markus CR, Olivier B, de Haan EHF. Whey protein rich in alpha-lactalbumin increases the ratio of plasma tryptophan to the sum of the other large neutral amino acids and improves cognitive performance in stress-vulnerable subjects. The American Journal of Clinical Nutrition. 2002;**75**(6):1051-1056

[65] Wurtman RJ, Wurtman JJ, Regan MM, McDermott JM, Tsay RH, Breu JJ. Effects of normal meals rich in carbohydrates or proteins on plasma tryptophan and tyrosine ratios. The American Journal of Clinical Nutrition. 2003;**77**(1):128-132

[66] Yokogoshi H, Wurtman RJ. Meal composition and plasma amino acid ratios: Effect of various proteins or carbohydrates, and of various protein concentrations. Metabolism. 1986;**35**(9):837-842

[67] Daniel PM, Love ER, Moorhouse SR, Pratt OE. The effect of insulin upon the influx of tryptophan into the brain of the rabbit. The Journal of Physiology. 1981;**312**:551-562

[68] Fukagawa NK, Minaker KL, Rowe JW, Young VR. Plasma tryptophan and total neutral amino acid levels in men: Influence of hyperinsulinemia and age. Metabolism. 1987;**36**(7):683-686

[69] Lipsett D, Madras BK, Wurtman RJ, Munro HN. Serum tryptophan level after carbohydrate ingestion: Selective decline in non-albumin-bound tryptophan coincident with reduction in serum free fatty acids. Life Sciences. 1973;**12**(Part 2):57-64

[70] Pardridge WM. Brain metabolism: A perspective from the blood-brain barrier. Physiological Reviews. 1983;**63**(4):1481-1535

[71] Trulson ME, Jacoby JH, MacKenzie RG. Streptozotocin-induced diabetes reduces brain serotonin synthesis in rats. Journal of Neurochemistry. 1986;**46**(4):1068-1072

[72] Manjarrez G, Herrera R, Leon M, Hernandez-R J. A low brain serotonergic neurotransmission in children with type 1 diabetes detected through the intensity dependence of auditory-evoked potentials. Diabetes Care. 2006;**29**(1):73-77

[73] Jain SK. Can tryptophan oxidation lead to lower tryptophan level in diabetes? A commentary on "Propagation of protein glycation damage involves modification of tryptophan residues via reactive oxygen species: inhibition by pyridoxamine". Free Radical Biology & Medicine. 2008;**44**(7):1273-1275

[74] Herrera-Marquez R, Hernandez-Rodriguez J, Medina-Serrano J, Boyzo-Montes de Oca A, Manjarrez-Gutierrez G. Association of metabolic syndrome with reduced central serotonergic activity. Metabolic Brain Disease. 2011;**26**(1):29-35

[75] Chen J, Ning C, Mu J, Li D, Ma Y, Meng X. Role of Wnt signaling pathways in type 2 diabetes mellitus. Molecular and Cellular Biochemistry. 2021;**476**(5):2219-2232

[76] Gürcü S, Girgin G, Yorulmaz G, Kılıçarslan B, Efe B, Baydar T. Neopterin and biopterin levels and tryptophan degradation in patients with diabetes. Scientific Reports. 2020;**10**(1):17025

[77] Jamnicky B, Slijepcević M, Hadzija M, Juretić D, Borcić O. Tryptophan content in serum and brain of long-term insulin-treated diabetic rats. Acta Diabetologica Latina. 1991;**28**(1):11-18

[78] MacKenzie RG, Trulson ME. Does insulin act directly on the brain to increase tryptophan levels? Journal of Neurochemistry. 1978;**30**(5):1205-1208

[79] MacKenzie RG, Trulson ME. Regional accumulation of tryptophan and serotonin metabolism following tryptophan loading in diabetic rats. Journal of Neurochemistry. 1978;**31**(1):157-160

[80] Curzon G, Fernando JC. Drugs altering insulin secretion: Effects on plasma and brain concentrations of aromatic amino acids and on brain 5-hydroxytryptamine turnover. British Journal of Pharmacology. 1977;**60**(3):401-408

[81] Newgard CB. Interplay between lipids and branched-chain amino acids in development of insulin resistance. Cell Metabolism. 2012;**15**(5):606-614

[82] Würtz P, Soininen P, Kangas AJ, Rönnemaa T, Lehtimäki T, Kähönen M, et al. Branched-chain and aromatic amino acids are predictors of insulin resistance in young adults. Diabetes Care. 2013;**36**(3):648-655

[83] Chen T, Zheng X, Ma X, Bao Y, Ni Y, Hu C, et al. Tryptophan predicts the risk for future type 2 diabetes. PLoS One. 2016;**11**(9):e0162192

[84] Vangipurapu J, Stancáková A, Smith U, Kuusisto J, Laakso M. Nine amino acids are associated with decreased insulin secretion and elevated glucose levels in a 7.4-year follow-up study of 5,181 Finnish men. Diabetes. 2019;**68**(6):1353-1358

[85] Giacco F, Brownlee M. Oxidative stress and diabetic complications. Circulation Research. 2010;**107**(9):1058-1070

[86] Brownlee M. The pathobiology of diabetic complications: A unifying mechanism. Diabetes. 2005;**54**(6):1615-1625

[87] Evans JL, Goldfine ID, Maddux BA, Grodsky GM. Are oxidative stress-activated signaling pathways mediators of insulin resistance and beta-cell dysfunction? Diabetes. 2003;**52**(1):1-8

[88] Chetyrkin SV, Mathis ME, Ham AJL, Hachey DL, Hudson BG, Voziyan PA. Propagation of protein glycation damage involves modification of tryptophan residues via reactive oxygen species: Inhibition by pyridoxamine. Free Radical Biology & Medicine. 2008;**44**(7):1276-1285

[89] Herrera R, Manjarrez G, Hernandez J. Inhibition and kinetic changes of brain tryptophan-5-hydroxylase during insulin-dependent diabetes mellitus in the rat. Nutritional Neuroscience. 2005;**8**(1):57-62

[90] Kanth VR, Lavanya K, Srinivas J, Raju TN. Elevated expression of indoleamine 2,3-dioxygenase (IDO) and accumulation of kynurenic acid in the pathogenesis of STZ-induced diabetic cataract in Wistar rats. Current Eye Research. 2009;**34**(4):274-281

[91] Oxenkrug G. Insulin resistance and dysregulation of tryptophan-kynurenine and kynurenine-nicotinamide adenine dinucleotide metabolic pathways. Molecular Neurobiology. 2013;**48**(2):294-301

[92] Friedman M, Cuq JL. Chemistry, analysis, nutritional value, and toxicology of tryptophan in food. A review. Journal of Agricultural and Food Chemistry. 1988;**36**(5):1079-1093

[93] Lin HV, Efanov AM, Fang X, Beavers LS, Wang X, Wang J, et al. GPR142 controls tryptophan-induced insulin and Incretin hormone secretion to improve glucose metabolism. PLoS One. 2016;**11**(6):e0157298

[94] Ueda Y, Iwakura H, Bando M, Doi A, Ariyasu H, Inaba H, et al. Differential role of GPR142 in tryptophan-mediated enhancement of insulin secretion in obese and lean mice. PLoS One. 2018;**13**(6):e0198762

[95] Heikkilä O, Lundbom N, Timonen M, Groop PH, Heikkinen S, Mäkimattila S. Hyperglycaemia is associated with changes in the regional concentrations of glucose and myo-inositol within the brain. Diabetologia. 2009;**52**(3):534-540

[96] Muriach M, Flores-Bellver M, Romero FJ, Barcia JM. Diabetes and the brain: Oxidative stress, inflammation, and autophagy. Oxidative Medicine and Cellular Longevity. 2014;**2014**:102158

[97] Shah GN, Morofuji Y, Banks WA, Price TO. High glucose-induced mitochondrial respiration and reactive oxygen species in mouse cerebral pericytes is reversed by pharmacological inhibition of mitochondrial carbonic anhydrases: Implications for cerebral microvascular disease in diabetes. Biochemical and Biophysical Research Communications. 2013;**440**(2):354-358

[98] Pandey SP, Singh HK, Prasad S. Alterations in hippocampal oxidative stress, expression of AMPA receptor GluR2 subunit and associated spatial memory loss by Bacopa monnieri extract (CDRI-08) in Streptozotocin-induced diabetes mellitus type 2 mice. PLoS One. 2015;**10**(7):e0131862

[99] Di Dalmazi G, Pagotto U, Pasquali R, Vicennati V. Glucocorticoids and type 2 diabetes: From physiology to pathology. Journal of Nutrition and Metabolism. 2012;**2012**:525093

[100] van Raalte DH, Ouwens DM, Diamant M. Novel insights into glucocorticoid-mediated diabetogenic effects: Towards expansion of therapeutic options? European Journal of Clinical Investigation. 2009;**39**(2):81-93

[101] Clark MS, Russo AF. Tissue-specific glucocorticoid regulation of tryptophan hydroxylase mRNA levels. Brain Research. Molecular Brain Research. 1997;**48**(2):346-354

[102] Ebou MH, Singh-Estivalet A, Launay JM, Callebert J, Tronche F, Ferré P, et al. Glucocorticoids inhibit basal and hormone-induced serotonin synthesis in pancreatic beta cells. PLoS One. 2016;**11**(2):e0149343

[103] Stetler C, Miller GE. Depression and hypothalamic-pituitary-adrenal activation: A quantitative summary of four decades of research. Psychosomatic Medicine. 2011;**73**(2):114-126

[104] Gupta D, Kurhe Y, Radhakrishnan M. Antidepressant effects of insulin in streptozotocin induced diabetic mice: Modulation of brain serotonin system. Physiology & Behavior. 2014;**22**(129):73-78

[105] Chen CC. Effect of the duration of streptozotocin-induced diabetes on turnover of central biogenic amines in mice. Neuroendocrinology. 1992;**56**(5):629-632

[106] Ludwig DS. The glycemic index: Physiological mechanisms relating to obesity, diabetes, and cardiovascular disease. Journal of the American Medical Association. 2002;**287**(18):2414-2423

[107] Ighodaro OM, Adeosun AM, Akinloye OA. Alloxan-induced diabetes, a common model for evaluating the glycemic-control potential of therapeutic compounds and plants extracts in experimental studies. Medicina. 2017;**53**(6):365-374

[108] Chen X, Yang W. Branched-chain amino acids and the association with type 2 diabetes. Journal of Diabetes Investigation. 2015;**6**(4):369-370

[109] Caballero B, Finer N, Wurtman RJ. Plasma amino acids and insulin levels in obesity: Response to carbohydrate intake and tryptophan supplements. Metabolism. 1988;**37**(7):672-676

[110] Brandacher G, Hoeller E, Fuchs D, Weiss HG. Chronic immune activation underlies morbid obesity: Is IDO a key player? Current Drug Metabolism. 2007;**8**(3):289-295

[111] Breum L, Rasmussen MH, Hilsted J, Fernstrom JD. Twenty-four-hour plasma tryptophan concentrations and ratios are below normal in obese subjects and are not normalized by substantial weight reduction. The American Journal of Clinical Nutrition. 2003;**77**(5):1112-1118

[112] Brandacher G, Winkler C, Aigner F, Schwelberger H, Schroecksnadel K, Margreiter R, et al. Bariatric surgery cannot prevent tryptophan depletion due to chronic immune activation in morbidly obese patients. Obesity Surgery. 2006;**16**(5):541-548

[113] Cangiano C, Laviano A, Del Ben M, Preziosa I, Angelico F, Cascino A, et al. Effects of oral 5-hydroxy-tryptophan on energy intake and macronutrient selection in non-insulin dependent diabetic patients. International Journal of Obesity and Related Metabolic Disorders. 1998;**22**(7):648-654

[114] Wurtman RJ, Wurtman JJ. Brain serotonin, carbohydrate-craving, obesity and depression. Obesity Research. 1995;**3**(Suppl. 4):477S-480S

[115] Amer A, Breu J, McDermott J, Wurtman RJ, Maher TJ. 5-Hydroxy-L-tryptophan suppresses food intake in food-deprived and stressed rats. Pharmacology, Biochemistry, and Behavior. 2004;**77**(1):137-143

[116] Gangwisch JE, Hale L, Garcia L, Malaspina D, Opler MG, Payne ME, et al. High glycemic index diet as a risk factor for depression: Analyses from the Women's health initiative. The American Journal of Clinical Nutrition. 2015;**102**(2):454-463

[117] McKeown NM, Meigs JB, Liu S, Saltzman E, Wilson PWF, Jacques PF. Carbohydrate nutrition, insulin resistance, and the prevalence of the metabolic syndrome in the Framingham offspring cohort. Diabetes Care. 2004;**27**(2):538-546

[118] Mwamburi DM, Liebson E, Folstein M, Bungay K, Tucker KL, Qiu WQ. Depression and Glycemic intake in the homebound elderly. Journal of Affective Disorders. 2011;**132**(1-2):94-98

[119] Møller SE. Serotonin, carbohydrates, and atypical depression. Pharmacology & Toxicology. 1992;**71**(Suppl. 1):61-71

[120] Wallin MS, Rissanen AM. Food and mood: Relationship between food, serotonin and affective disorders. Acta Psychiatrica Scandinavica. Supplementum. 1994;**377**:36-40

[121] Andrews PW, Thomson JA, Amstadter A, Neale MC. Primum non Nocere: An evolutionary analysis of whether antidepressants do more harm than good. Frontiers in Psychology. 2012;**24**(3):117

[122] Isaac R, Boura-Halfon S, Gurevitch D, Shainskaya A, Levkovitz Y, Zick Y. Selective serotonin reuptake inhibitors (SSRIs) inhibit insulin secretion and action in pancreatic β cells. The Journal of Biological Chemistry. 2013;**288**(8):5682-5693

[123] Yao S, Li J, Fan X, Liu Q, Lian J. The effect of selective serotonin re-uptake inhibitors on risk of type II diabetes mellitus and acute pancreatitis: A meta-analysis. Bioscience Reports. 2018;**38**(5):BSR20180967

[124] Levkovitz Y, Ben-Shushan G, Hershkovitz A, Isaac R, Gil-Ad I, Shvartsman D, et al. Antidepressants induce cellular insulin resistance by activation of IRS-1 kinases. Molecular and Cellular Neurosciences. 2007;**36**(3):305-312

[125] Pan A, Sun Q, Okereke OI, Rexrode KM, Rubin RR, Lucas M, et al. Use of antidepressant medication and risk of type 2 diabetes: Results from three cohorts of US adults. Diabetologia. 2012;**55**(1):63-72

[126] Khoza S, Barner JC, Bohman TM, Rascati K, Lawson K, Wilson JP. Use of antidepressant agents and the risk of type 2 diabetes. European Journal of Clinical Pharmacology. 2012;**68**(9):1295-1302

[127] Burcu M, Zito JM, Safer DJ, Magder LS, dosReis S, Shaya FT, et al. Association of Antidepressant Medications with Incident Type 2 diabetes among Medicaid-insured youths. JAMA Pediatrics. 2017;**171**(12):1200-1207

[128] Miidera H, Enomoto M, Kitamura S, Tachimori H, Mishima K. Association between the use of antidepressants and the risk of type 2 diabetes: A large, population-based cohort study in Japan. Diabetes Care. 2020;**43**(4):885-893

[129] Spillmann MK, Van der Does AJ, Rankin MA, Vuolo RD, Alpert JE, Nierenberg AA, et al. Tryptophan depletion in SSRI-recovered depressed outpatients. Psychopharmacology. 2001;**155**(2):123-127

[130] Argyropoulos SV, Hood SD, Adrover M, Bell CJ, Rich AS, Nash JR, et al. Tryptophan depletion reverses the therapeutic effect of selective serotonin reuptake inhibitors in social anxiety disorder. Biological Psychiatry. 2004;**56**(7):503-509

[131] Hari J. Lost Connections: Uncovering the Real Causes of Depression - and the Unexpected Solutions [Internet]. USA: Bloomsbury Publishing; 2018 [cited 2021 Sep 8]. Available from: https://librarysearch.library.utoronto.ca/discovery/fulldisplay/cdi_proquest_ebookcentral_EBC5235822/01UTORONTO_INST:UTORONTO

[132] Baker S. Complex Adaptive Systems (CAS) Approach to Biomedicine & Public Health [Internet]. 2021 [cited 2023 Nov 8]. Available from: https://anatomisebiostats.com/biostatistics-blog/complex-adaptive-systems-cas-approach-to-biomedicine-public-health/

[133] Wilson T, Holt T. Complexity science: Complexity and clinical care. BMJ : British Medical Journal. 2001;**323**(7314):685

[134] Ridder DD, Stöckl T, To WT, Langguth B, Vanneste S. Chapter 7 - Noninvasive transcranial magnetic and electrical stimulation: Working mechanisms. In: Evans JR, Turner RP, editors. Rhythmic Stimulation Procedures in Neuromodulation. London: Academic Press; 2017. pp. 193-223. Available from: https://www.sciencedirect.com/science/article/pii/B9780128037263000079

[135] Kay J, Schneider E. Embracing complexity: The challenge of the ecosystem approach. Alternatives. 1994;**20**:32-39

[136] Loring PA. Threshold concepts and sustainability: Features of a contested paradigm. Klenk NL, editor. FACETS. 2020;**5**(1):182-199

[137] Jayasinghe S. Complexity science to conceptualize health and disease: Is it relevant to clinical medicine? Mayo Clinic Proceedings. 2012;**87**(4):314-319

[138] Loring PA. A vernacular for living systems: Alternative framings for the future of food. Futures. 2023;**1**(154):103276

[139] Du Plessis C, Brandon P. An ecological worldview as basis for a regenerative sustainability paradigm for the built environment. Journal of Cleaner Production. 2015;**109**:53-61

[140] Carey AL, Kingwell BA. Reducing peripheral serotonin turns up the heat in brown fat. Nature Medicine. 2015;**21**(2):114-116

[141] Crane JD, Palanivel R, Mottillo EP, Bujak AL, Wang H, Ford RJ, et al. Inhibiting peripheral serotonin synthesis reduces obesity and metabolic dysfunction by promoting brown adipose tissue thermogenesis. Nature Medicine. 2015;**21**(2):166-172

[142] Paulmann N, Grohmann M, Voigt JP, Bert B, Vowinckel J, Bader M, et al. Intracellular serotonin modulates insulin secretion from pancreatic beta-cells by protein serotonylation. PLoS Biology. 2009;**7**(10):e1000229

[143] Lam DD, Heisler LK. Serotonin and energy balance: Molecular mechanisms and implications for type 2 diabetes. Expert Reviews in Molecular Medicine. 2007;**9**(5):1-24

[144] Inubushi T, Kamemura N, Oda M, Sakurai J, Nakaya Y, Harada N, et al. L-tryptophan suppresses rise in blood glucose and preserves insulin secretion in type-2 diabetes mellitus rats. Journal of Nutritional Science and Vitaminology (Tokyo). 2012;**58**(6):415-422

[145] Grady CL, Siebner HR, Hornboll B, Macoveanu J, Paulson OB, Knudsen GM. Acute pharmacologically induced shifts in serotonin availability abolish emotion-selective responses to negative face emotions in distinct brain networks. European Neuropsychopharmacology. 2013;**23**(5):368-378

[146] Cools R, Robinson OJ, Sahakian B. Acute tryptophan depletion in healthy volunteers enhances punishment prediction but does not affect reward prediction. Neuropsychopharmacology. 2008;**33**(9):2291-2299

[147] Robinson OJ, Charney DR, Overstreet C, Vytal K, Grillon C. The adaptive threat bias in anxiety: Amygdala-dorsomedial prefrontal cortex coupling and aversive amplification. NeuroImage. 2012;**60**(1):523-529

[148] Bengart P, Gruendler T, Vogt B. Acute tryptophan depletion in healthy

subjects increases preferences for negative reciprocity. PLoS One. 2021;**16**(3):e0249339

[149] Badcock PB, Davey CG, Whittle S, Allen NB, Friston KJ. The depressed brain: An evolutionary systems theory. Trends in Cognitive Sciences. 2017;**21**(3):182-194

[150] Russo SJ, Murrough JW, Han MH, Charney DS, Nestler EJ. Neurobiology of resilience. Nature Neuroscience. 2012;**15**(11):1475-1484

[151] Durisko Z, Mulsant BH, Andrews PW. An adaptationist perspective on the etiology of depression. Journal of Affective Disorders. 2015;**172**:315-323

[152] Andrews PW, Thomson JA. The bright side of being blue: Depression as an adaptation for analyzing complex problems. Psychological Review. 2009;**116**(3):620-654

[153] Booij L, Van der Does W, Benkelfat C, Bremner JD, Cowen PJ, Fava M, et al. Predictors of mood response to acute tryptophan depletion. A reanalysis. Neuropsychopharmacology. 2002;**27**(5):852-861

[154] Smith KA, Fairburn CG, Cowen PJ. Relapse of depression after rapid depletion of tryptophan. Lancet. 1997;**349**(9056):915-919

[155] Miller HE, Deakin JF, Anderson IM. Effect of acute tryptophan depletion on CO2-induced anxiety in patients with panic disorder and normal volunteers. The British Journal of Psychiatry. 2000;**176**:182-188

[156] Maletic V, Robinson M, Oakes T, Iyengar S, Ball SG, Russell J. Neurobiology of depression: An integrated view of key findings. International Journal of Clinical Practice. 2007;**61**(12):2030-2040

[157] Slavich GM, Irwin MR. From stress to inflammation and major depressive disorder: A social signal transduction theory of depression. Psychological Bulletin. 2014;**140**(3):774-815

[158] Palego L, Betti L, Rossi A, Giannaccini G. Tryptophan biochemistry: Structural, nutritional, metabolic, and medical aspects in humans. Journal of Amino Acids. 2016;**2016**:8952520

[159] Zhou J, Li L, Tang S, Cao X, Li Z, Li W, et al. Effects of serotonin depletion on the hippocampal GR/MR and BDNF expression during the stress adaptation. Behavioural Brain Research. 2008;**195**(1):129-138

[160] Chung KK, Martinez M, Herbert J. Central serotonin depletion modulates the behavioural, endocrine and physiological responses to repeated social stress and subsequent c-fos expression in the brains of male rats. Neuroscience. 1999;**92**(2):613-625

[161] Iacoviello BM, Charney DS. Psychosocial facets of resilience: Implications for preventing posttrauma psychopathology, treating trauma survivors, and enhancing community resilience. European Journal of Psychotraumatology. 2014;**2014**:5 [cited 2019 Mar 17]. Available from: https://www.ncbi.nlm.nih.gov/pmc/articles/PMC4185137/

[162] Osei K, Rhinesmith S, Gaillard T, Schuster D. Impaired insulin sensitivity, insulin secretion, and glucose effectiveness predict future development of impaired glucose tolerance and type 2 diabetes in pre-diabetic African Americans: Implications for primary diabetes prevention. Diabetes Care. 2004;**27**(6):1439-1446

[163] Tabák AG, Herder C, Rathmann W, Brunner EJ, Kivimäki M. Prediabetes: A high-risk state for developing diabetes. Lancet. 2012;**379**(9833):2279-2290

Chapter 2

Serotonin: The Link between Gut Microbiome and Brain

Mushtaq Margoob, Shazia Kouser and Neelofer Jan

Abstract

Serotonin, as a neurotransmitter plays a key role in regulating mood, sleep, appetite, and various physiological processes. Serotonin is closely linked to the microbiome-gut-brain axis, which is a bidirectional communication between the gut and the brain facilitated by the gut microbiome which consists of trillions of microorganisms that inhabit the digestive tract. This connection is a growing area of research and serotonin produced in the gut is being investigated for its potential impact on human personality, mood, and overall health. Microbiome influences serotonin production, serotonin precursor metabolism, serotonin reuptake, and immune system modulation. A balanced microbiome is crucial for regulating homeostasis and stress response and altered gut microbiota composition has been linked to depression, anxiety, bipolar, schizophrenia, stress-related, and autism spectrum disorders. Microbiome-based interventions might help to regulate the immune response, neuroprotection, and neuroplasticity to reduce neuroinflammation and thus prove crucial to modifying the course of major depressive, bipolar, and related disorders where inflammation is evidenced to lead to the progression of illnesses. Microbiome-based interventions such as probiotic supplementation influence the production of neuroactive compounds and have the potential to bridge the treatment gap for Parkinson's disease, multiple sclerosis, and Alzheimer's disease and might prove to be a turning point for the treatment of obesity-associated systemic low-level inflammation, whether psychotropic medication related or otherwise. The gut microbiome offers a novel possibility to employ manipulation of the gut microbiota as a non-invasive measure in health and disease, especially at a time when the clinical field of forthcoming psychotropics looks exhausted.

Keywords: serotonin, gut-brain axis, microbiome, probiotics, stress

1. Introduction

Serotonin is a biologically active amine that serves the dual functions of neurotransmitters and hormones by exerting a wide range of physiological and pathological effects through nearly a dozen receptors classified into seven families [1]. Serotonin also known as enteramine or 5-hydroxytryptamine (5-HT) was successfully extracted and purified by Rapport and colleagues in 1948 [2].

1.1 Serotonin in the brain

1.1.1 Anatomy

The distribution of serotonin in CNS consists of the following:

1. Raphe Nuclei (Latin, meaning "midline"): Serotonin within the CNS is almost exclusively produced in neurons originating in the raphe nuclei which are collections of neurons, with poorly defined cytoarchitectonic limits localized to the periaqueductal gray, also known as central gray, and the surrounding reticular formation in the brain.

2. Long and extensively branched axonal processes called projections. Serotonergic neurons from the raphe nuclei project widely throughout the CNS and form classical chemical synapses as well as such synapses that contribute to the so-called paracrine or volume transmission, and this has led to the suggestion that serotonin exerts a major modulatory role throughout the CNS.

3. Multiple cortical and limbic target regions: The raphe nuclei provide projections to the cortex, and many forebrain limbic structures such as the hippocampus and medulla. Projections to the dorsal, intermediate, and ventral columns in the spinal cord regulate pain perception at the level of the dorsal horn. Serotonergic terminals in the cortex are less organized than the noradrenergic cortical projections, however, the two systems are co-localized in most limbic areas of the brain and this might explain the major involvement of these transmitters in the affective disorders [3–5].

1.1.2 Neurochemistry

Serotonin is synthesized from the essential amino acid L-tryptophan (Trp) which is primarily obtained from dietary sources [6]. After absorption about 85% of tryptophan is bound to plasmatic albumin protein and only 10–20% is unbound and able to cross blood–brain barrier (BBB) and hence available for 5-HT synthesis in the brain. Tryptophan released from plasma proteins becomes available for incorporation into proteins [7] as this is the principal role of tryptophan in the human body [8]. The second most prevalent metabolic pathway of tryptophan, the kynurenine pathway, accounts for the catabolism of approximately 99% of ingested tryptophan not used for protein synthesis and has importance in generating cellular energy in the form of nicotinamide adenine dinucleotide (NAD). The kynurenine pathway produces other pro and antioxidant molecules of neurobiological importance namely, kynurenine, kynurenic acid, and quinolinic acid (QUIN) [9, 10]. Synthesis of B6 and B12 vitamins, required as co-factors for kynurine pathway enzymes are dependent on gut microbiome activity [11]. The third tryptophan metabolic pathway that leads to the synthesis of 5-HT in the periphery (e.g. in blood platelets and the enterochromaffin cells of the gastrointestinal tract) or in the nerve endings in brain is relatively minor. It is estimated that 95% of mammalian serotonin is found within the gastrointestinal tract and while 3% of dietary tryptophan is used for serotonin synthesis throughout the body only 1% of dietary tryptophan is used for the synthesis of this broad-impact neurotransmitter and neuromodulator in the brain. Melatonin and tryptamine are other by-products of the tryptophan/serotonin pathway [12]. In serotonin synthesis from tryptophan, the first step is catalyzed by the enzyme tryptophan hydroxylase which exists in two isoforms tryptophan hydroxylase 1 (*Tph1*) and tryptophan hydroxylase 2 (*Tph 2*) and convert tryptophan to 5-hydroxytryptophan. 5-Hydroxytryptophan is then converted by the aromatic amino acid decarboxylase to 5-hydroxytryptamine [13]. After synthesis serotonin (5HT) is transported into synaptic vesicles using vesicular monoamine transporter 2 (VMAT2) to protect serotonin from enzymatic breakdown. Once released from the presynaptic terminal serotonin acts on various presynaptic and postsynaptic receptors which are also targets of numerous drugs. The action of serotonin on its receptors at the synapses is terminated mainly by an active reuptake process mediated by the serotonin transporter (SERT), a sodium and chloride-dependent neurotransmitter transporter [14, 15].

1.2 Role of serotonin in mood regulation and emotional well-being

Speculations about the role of monoamines in affective states began with the serendipitous discovery in the late 1950s that members of two structurally unrelated classes of compounds monoamine oxidase inhibitors (MOAIs) and tricyclic antidepressants (TCAs) were effective in treating severe depression [16]. In the most basic form, Monoamine Theories postulate that depression is related to decreased levels of centrally available monoamines, typically either the catecholamine, noradrenaline (norepinephrine in the United States), or the indoleamine, serotonin [17]. Monoamine theories later evolved into monoamine receptor theories, which associate depression with lesions at the level of monoamine receptors [18]. With the advancement in neurosciences, even receptor theories have come under scrutiny, and contemporary literature has expanded these theories to the non-mutually exclusive neurotrophic and neurogenesis hypotheses which are closely related to a third entity called neurotrophic hypothesis. These hypotheses were proposed as the monoamine theory of depression was too simplistic to explain several conundrums which evidenced that monoamine deficiency cannot be the sole cause of depression [19]. The Neuroplasticity hypothesis postulates that depression may result from environmental contingencies like adverse life experiences that cause neuronal architectural changes and resultant defects in brain processing, the pathophysiology of which is strongly linked to impairments in serotonin (5-HT) neurotransmission [20]. Neurogenesis research suggests that very sophisticated transport systems can allow freshly produced neurons (from lateral subventricular zone (SVZ) and the dentate gyrus (DG) of the hippocampus and the olfactory bulb), called "neuroblasts," to be migrated long distances across the brain to help regenerate damaged areas or regions which are experiencing neural dilapidation and thus neurogenesis is thought to be important for maintaining brain health. Antidepressants might improve neurogenesis in the hippocampus through activation of the 5-HT1A receptor [21, 22]. The neurotrophic hypothesis of depression posits that major depressive disorder (MDD) is caused partly by decreases in neurotrophic factors such as brain-derived neurotrophic factor (BDNF) and their restoration is critical for the therapeutic efficacy of antidepressant treatment [23]. Emotional and cognitive deficits in bipolar and related mood disorders are linked to changes in neuroplasticity, cell resilience, and connectivity with BDNF as an important contributor to the neuroplasticity changes described among bipolar disorder patients. Thus, evidence from differential lines of research converges to serotonin signaling and 5-HT receptors being involved in regulating the levels of both neurotrophic factors and adult hippocampal neurogenesis thereby mediating a pivotal role in neuroplasticity in both normal and neuropsychiatric conditions

[24–26]. Further, the neuroplasticity theory is not considered exclusive for MDD, and the mechanisms of alterations in neuroplasticity accounting for the significantly different symptomatology of schizophrenia [27] and bipolar disorders [28] are being investigated.

1.3 Other roles of serotonin

1.3.1 Cognition

The role of serotonin in human cognition has been investigated, especially in the context of the growing notion of memory deficits in neuropsychiatric disorders like posttraumatic stress disorder, schizophrenia, depression, and Parkinson's disease [29]. 5-HT6 antagonists are expected to be effective against learning impairment from anticholinergic and antiglutamatergic models of dementia. It is supposed that the procognitive activity of some marketed antidepressants (Vortioxetine) and antipsychotics (Lurasidone) is caused by potent 5-HT7R affinity, and both 5-HT6 and 5-HT7 receptors represent interesting targets in the search for innovative therapies of AD [30]. Evidence is mounting for the role of 5-HT in human cognition and normalizing 5-HT activity in depression and Alzheimer's disease (AD) may have specific beneficial effects on cognition, independent of a general relief of mood symptoms, however, as of now, data is not sufficient to comment on emergent use of 5-HT targeting drugs as potential cognition enhancers [31].

1.3.2 Appetite

Brainstem-derived serotonin influences cognitive functions including eating behaviors and regulates homeostatic functions of bone remodeling, appetite, and energy expenditure. Gut-derived serotonin, on the other hand, plays a critical role in feeding activity. Serotonin thus acts as a hormone when made in the gut and a neurotransmitter when made in the brain [32, 33]. Although social and psychological aspects of eating are powerful influences that are independent of or only partially dependent on the physiologic control mechanisms, at the receptor level there is evidence that the 5-HT1B and 5-HT2C receptors are involved in mediating the effects of serotonergic drugs on food intake. Appetite suppression appears to be associated with agonist action at 5-HT2C receptors in the central nervous system. 5-HT2C agonist, lorcaserin, is approved by the FDA for use as a weight-loss medication in monotherapy. While no available pharmacologic therapy has succeeded in maintaining a weight loss of over 10% for 1 year, bariatric (weight-reducing) surgery readily achieves a sustained weight loss of 10–40%, and surgery that bypasses the stomach and upper small intestine rapidly reverses some aspects of the metabolic syndrome. Gastrointestinal flora also influence energy expenditure, and research suggests that altering the microbiome can lead to weight gain or loss [1, 34].

1.3.3 Sleep

Serotonin along with fast-acting non-monoaminergic neurotransmitters (glutamate and GABA) and other monoaminergic neurotransmitters plays a crucial role in regulating the sleep–wake cycle. Earlier it was thought that serotonin might help to produce NREM and possibly REM sleep, however, more recent work indicates that serotonin generally promotes wakefulness and suppresses REM sleep. The role of serotonin in sleep is, however, not straightforward. On one hand, serotonin inhibits the wake-promoting cholinergic neurons and serves as a precursor of melatonin in the pineal gland where 5-HT is O-methylated to form melatonin. In humans, melatonin plays a significant role in both inducing and maintaining nocturnal sleep and drives the circadian rhythm which in turn regulates the sleep–wake cycle, and endocrine, immune and neurotransmitter rhythmicity [35]. On the other hand, the firing rates of dorsal raphe neurons and extracellular 5-HT levels are highest during wakefulness, much lower during NREM sleep, and lowest during REM sleep. This wake-promoting role of serotonin is further evidenced by agonists of the 5-HT 1A, 5-HT 1B, 5-HT2, or 5-HT3 receptors that increase wakefulness and 5-HT2 receptor blockers such as ritanserin or agomelatine that promote NREM sleep [36].

1.3.4 Pain

Advances in basic sciences, clinical research and now neuroimaging have established that central sensitization and alterations in neuroplasticity induced by the enhancement of descending pain facilitation and/or the impairment of descending pain inhibition underlie many chronic pain conditions. The descending serotonergic neurons in the raphe nuclei target receptors along the descending pain circuits and exert either pro- or antinociceptive effects, thus, serotonin has a definite role in the pathogenesis of chronic pain conditions like chronic primary pain (CPP), inflammatory bowel disease (IBD), Fibromyalgia syndrome (FMS), etc. [37]. Antidepressants like TCAs, SNRIs, and SSRIs influence the descending pain modulation system by increasing 5-HT at the synaptic junction [5, 38].

1.3.5 Neurological diseases

Migraine, epilepsy, Parkinson's disease (PD), multiple sclerosis (MS), ALS, and neuropsychiatric disorders (ADHD, ASD) are connected to abnormal 5-HT synthesis and metabolism as the efficiency of 5-HT metabolism is changed in neurodegeneration. Patients with amyotrophic lateral sclerosis (ALS) and multiple sclerosis (MS) present with reduced plasma and CSF levels of tryptophan and subsequently a decreased 5-HT synthesis. In MS, 5-HT synthesis is decreased because of the overactivation of the kynurenine pathway, which drives Tryptophan away from 5-HT synthesis [39]. Migraine is caused by a decreased level of platelet serotonin and its metabolite N-acetylserotonin (NAS) that activate trigeminovascular system (TGVS) and lead to cortical spreading depression (CSD) during an acute attack of migraine. Triptans, acting via 5-HT 1B receptor and serotonin activity at the 5-HT1F receptor on neuronal synapses inhibiting the release of calcitonin gene-related peptide (CGRP) are disease-specific treatments of migraine. It has long been known that serotonin inhibits epileptic activity and because of its crucial role in influencing seizures, regulating sleep and wakefulness, arousal, circadian rhythms, breathing, and cardiac activity, serotonin (5-HT) has been implicated in the pathophysiology of sudden unexpected death in epilepsy (SUDEP) [40]. Apart from other actions, CBZ and VPA release serotonin and LTG inhibits serotonin uptake, however, only a few AEDs, such as the recently approved fenfluramine, act via 5-HT receptors [41]. Progressive dopaminergic denervation is the cardinal pathology in Parkinson's disease, however, several lines of evidence suggest that a progressive and non-linear loss of serotonergic terminals which is not related to disease duration, disability or dopamine replacement therapy takes place in Parkinson's disease though at a slower rate. Human PET studies indicate that striatal serotonergic terminals contribute to Levodopa-induced dyskinesias (LIDs) via aberrant processing wherein serotonergic neurons take up, convert exogenous Levodopa into dopamine, and release dopamine as a false neurotransmitter in the denervated striatum of PD patients with LIDs. This study also speculates the development of selective serotonin receptor type 1A agonists for use as antidyskinetic agents in PD [42]. In clinical studies, the nonselective 5-HT 1A agonist buspirone reduced LID without worsening parkinsonian disability. 5-HT 2C receptor antagonism is a potential mechanism whereby clozapine and quetiapine can reduce LID [42, 43]. Abnormalities in SERT and MAO-A activity in various brain regions have been found to be associated with impulsivity and aggressive tendencies in ADHD. 5-HT deficiency leads to a failure of 5-HT-mediated inhibition of aggressive behavior in adults as well as children and decreased levels of 5-HT and its metabolite 5-HIAA, in the blood, urine, and CSF in individuals with ADHD compared with healthy controls. Although precise pathomechanism of ASD has not been elucidated, hyperserotonemia is present in approximately 30% of patients of ASD. One of the consequences of hyperserotonemia is increased catabolism of 5-HT [39].

2. Serotonin in the gut

Nearly 95% of the body's content of serotonin is found in GIT and only 5% is found in the brain. Within the gut, about 90% of serotonin is in EC cells and about 10% is found in enteric neurons, pancreatic cells and mast cells. Enterochromaffin (EC) cells are excitable, serotonergic neuroendocrine cells located throughout the length of the lining of the gastrointestinal tract. EC cells synthesize 5-HT, in the presence of some cofactors, such as vitamin B6, vitamin B3, and magnesium, from its precursor L-tryptophan in a reaction catalyzed by the enzyme tryptophan hydroxylase, which exists in two isoforms (Tph1 and Tph2). Tph1 is mainly present in EC whereas Tph2 is found in CNS and enteric neurons [44]. EC cells release 5-HT in a regulated manner in response to various mechanical and chemical stimuli. 5-HT thus released from EC cells reaches the blood, surrounding tissues, and gut lumen. Once released, 5-HT is transported into surrounding epithelial cells and platelets by the serotonin reuptake transporter (SERT) and degraded to 5-hydroxyindoleacetic acid (5-HIAA). Platelets are a major source of peripheral 5-HT as they store the 5-HT synthesized by EC cells in the gut and are always present in the circulation. Five (5-HTR1, 5-HTR2, 5-HTR3, 5-HTR4, and 5-HTR7) of the seven 5-HT receptor (5-HTR) families are expressed in the gut smooth muscle, enteric neurons, enterocytes, and immune cells through which serotonin mediates various secretomotor and sensory functions such as nausea, vomiting, intestinal fluid and mucus secretion and peristaltic movement [45].

2.1 Functions of serotonin in the gastrointestinal tract

1. Serotonin as a regulator of gut motility: 5-HT plays a crucial role in the generation of peristaltic reflexes, segmentation, and mucosal stimulation in response to food intake, under normal circumstances and in disorders of GI tract associated with the alteration of motility and sensation like the irritable bowel syndrome (IBS) [44].

2. Serotonin in fluid and mucus secretion: 5-HT inhibits gastric acidity by increasing the gastric mucus and fluid secretion. Mucus in GIT acts as a physical barrier for microorganisms, diffusion of toxins, and as an antioxidant [46].

3. Pain and anxiety: EC cell activation produces persistent visceral hypersensitivity in response to gut distension, even in the absence of inflammation.

4. Serotonin in immune cell function: Serotonin receptors are expressed by nearly all innate immune cells, such as the Langerhans cells or the immature dendrite cells (DCs) in the skin and other epithelial tissues lining the nose, lungs, stomach and intestine, monocytes, mast cells and eosinophils. 5-HT, along with other platelet-derived factors, plays a crucial role in the recruitment of these cells at the site of acute inflammation [47].

5. Serotonin in inflammation: Serotonin evokes divergent pro as well as anti-inflammatory actions and while 5-HTR4-mediated anti-inflammatory action predominates in the basal or normal conditions, 5-HTR7-mediated pro-inflammatory signals predominate under pathological conditions. Inflammatory bowel diseases (IBD) viz., Crohn's disease (CD) and ulcerative colitis (UC) are believed to result from an abnormal response to self-antigens or the gut resident microbiota and are characterized by activation of both innate and adaptive immune systems in response to enhanced cytokine production in GIT [48].

6. Angiogenesis: 5-HT release from platelets stimulates angiogenesis in many physiological processes, such as organ development, reproduction, wound healing and pathological conditions like IBD, diabetic retinopathy, rheumatoid arthritis, age-related macular degeneration, and tumor growth and metastasis [49].

3. The gut microbiome

Conventionally brain has been considered sealed from microbial influence unless infection occurs, however, evidence in favor of gut microbiota interplays with different systems ultimately impacting the brain has accumulated over the past few decades to the extent that the gut microbiome has earned the name "second brain" from some authors.

3.1 Composition and diversity of gut microbiome

The term microbiome pertains to the community of microorganisms, their structure, activity, metabolites and mobile genetic elements while microbiota is a collection of microbial communities associated with a habitat [11]. This collection of bacteria, viruses, fungi, protozoans, and archaea found in an individual constitutes about 1–3% of total human body mass. Majority of the human microbiome is comprised of bacteria, about 100 trillion bacteria from 500 to 1000 different species, with varying diversity adding over eight million genes to the human genome. The microbiota colonize predominantly the human intestine and to a lesser extent the airways and the skin surface. A simplified taxonomic composition of gut microbiota is presented in **Table 1**. Most of these microorganisms belong to the phyla Firmicutes and Bacteroidetes [51].

3.2 Factors influencing the composition of the microbiome

Gut microbiome is a dynamic with many variables influencing its composition.

Phylum	Examples
1. Actinobacteria	*Bifidobacterum longum* *Bifidobacterium bifidum*
2. Firmicutes	*Faecalibacteruim prausnitzii* *Clostridium* spp. *Roseburia intestinalis* *Runinococcus feacis* *Dialister invisus* *Lactobacillus renteri* *Enterococcus faecium* *Staphylococcus leei*
3. Bacteriodetes	*Bacteriodes fragilis* *Bacteriodes vulgaris* *Bacteriodes uniformis* *Parabacteriodes diastaoms* *Alistipes finegoldii* *Prevotella* spp.
4. Proteobacteria	*Escherichia coli* *Shigella flexineria* *Desulforibrio intestinalis* *Bilophila* *Wadsworthia*
5. Fusobacteria	*H. Pylori* *Fusobacterium nucleatum*
6. Verrucomicrobia	*Akkermansia muciniphila*

Table 1.
Simplified taxonomic classification of gut microbial composition [50].

3.2.1 Genetics

Twin studies have revealed similar microbiome composition in monozygotic twins and this similarity has been seen to be more in monozygotic twins and dizygotic twins than in other family members [52]. Genetic animal models of 5-HTT deficiency have revealed the presence of altered microbial composition in 5HTT knockout mice such as they had predominance of pathobionts [53].

3.2.2 Early life factors

Microbes colonize the various sites from the first days of life, reach high numbers immediately after birth, and gradually evolve and diversify with the growth of the individual to outnumber somatic cells by a number of ten. Microbiota are shaped in the first few years of life by gut maturation developing from enterotypes, which are functionally harmonious clusters of bacteria that characterize individuals and are regrouped by functions. The first 2 years of life including the intrauterine period seem to represent the most critical time for microbiome modulation. Other factors modulating this composition of microbiota in early life are birth gestational age, type of delivery, methods of milk feeding, weaning period, maternal diet/weight, pro and prebiotic use, early antibiotic exposure, timing and type of complementary feeding, lifestyle, dietary and cultural habits [54–56].

3.2.3 Gut permeability and inflammation

Gut epithelium serves a protective and structural role in the human body and when this barrier is compromised, the so-called "leaky gut" is associated with pathological conditions that activate gut pain sensory pathways and dysregulate the enteric nervous system. The stress of varying types can impact the developmental trajectory of intestinal barrier by causing significant perturbations in gut permeability as well as gut microbiome [57, 58] and maternal separation has been shown to cause such a shift in the microbial composition in a drastic way [59, 60].

3.2.4 Body mass index (BMI) classes and exercise frequency

Gut microbiota variations are correlated with obesity, anorexia nervosa and exercise as a form of environmental enrichment has been shown to impact the gut microbiome in a positive way [61, 62].

3.2.5 Aging

Microbial changes that occur with Aging have been grouped into two categories; those associated with healthy aging and pathobionts associated with ill health in aging [63].

3.3 Gut-brain axis

It is a complex, firmly established, bidirectional network that connects the microbiota, enteric and central nervous system. The microbiome gut-brain axis (MGBA) can be modulated by endocrine, neural and immune pathways in a bottom-up or top-down approach with multiple feedback loops regulating this network. In top-down approach, the brain uses these mechanisms to influence the composition of microbiota in gut. In the bottom-up approach microbiome signals brain through immune regulation by the production of cytokines and through production of neurotransmitters and neuroactive metabolites like short-chain fatty acids (SCFAs) [11, 64].

3.4 Pathways of the gut-brain axis

3.4.1 Neurologic pathway

The vagus nerve tonically transmits information from the viscera to the brain and vice versa and is considered to be the fastest and most direct way for the microbiota to influence the brain. Specific bacteria within the gut microbiota utilize the vagus nerve to communicate with the brain to alter certain neurocircuits by affecting primary afferent neuronal excitability. Ablation of gut-related vagal communication between lower GI tract and brain, as evidenced by animal studies and surgical procedures like gastrectomies, resulting in changes in adult neurogenesis, stress reactivity, cognition, and increased occurrences of psychiatric-related disorders has been recognized for long [65, 66].

3.4.2 Endocrine pathway

The hypothalamus pituitary adrenal (HPA) axis regulates cortisol secretion in response to stress by directly affecting immune cells and release of cytokines systemically as well as locally in the gut. Cortisol affects gut permeability, its barrier function as well as composition of gut microbiota. Gut microbiome, in a bottom-up fashion, influences the release of cytokines and other immune mediators like interferon-gamma. Serotonin plays a crucial role in recruitment of innate immune cells in response to this cytokine release during time of dysbiosis. Gut microbiome also influences the release of neuropeptides like galanin, leptin and neuropeptide Y (NPY) from enteroendocrine cells which reach the systemic circulation and bind receptors on immune cells and vagus nerve terminals thereby enabling indirect gut-brain communication [67–69]. Another mechanism of microbiome-gut-brain crosstalk is through tryptophan and its metabolites such as 5-hydroxyindoleacetic acid (5-HIAA). The gut microbiota can alter concentrations of kynurine and disruption of this metabolic pathway has been linked to both GI and brain disorders.

3.4.3 Metabolic/humoral pathway

Bacterial metabolites like short-chain fatty acids [SCFAs] and lipopolysaccharides, which are produced by their fermentation of dietary carbohydrates are important humoral influencers. These metabolites affect the nutrition of the enterocytes, possess hormone-like activity, stimulate the sympathetic nerves of gut and also have immunomodulatory properties. SCFAs also regulate microglial homeostasis which in turn affects brain development, brain tissue homeostasis, and behavior [70].

3.5 Impact of the gut microbiota on serotonin levels in gut and brain

Serotonin is directly synthesized by commensal bacteria from the colonic luminal tryptophan and serotonin biosynthesis is promoted in the colonic ECs by spore-forming bacteria. Gut microbiota promotes enteric 5-HT production through SCFAs as well as phenolic and indolic compounds derived from microbes. Microbiota affects the central serotonin levels through many pathways. It influences availability of the peripheral tryptophan by affecting the metabolism of the gut luminal tryptophan thereby altering the central tryptophan levels and hence the central serotonin levels. The microbial metabolites e.g., SCFAs, especially butyrate have been reported to increase brain serotonin concentration. In addition, inflammatory stimuli, such as LPS, the major components of the outer membrane of Gram-negative bacteria, have been postulated to affect the kynurenine pathway thereby diverting tryptophan away from serotonin synthesis. In gut dysbiosis, as demonstrated by antibiotic studies, central levels of serotonin and its precursor tryptophan has been seen to be reduced and these are posited to be due to increased serotonin metabolism as reflected by increased SERT and MAO expression in the hypothalami of piglets. The microbiome also regulates serotonin transporter (SERT) expression by gut bacteria via posttranslational and transcriptional mechanisms, alterations in SERT surface levels, and epigenetic or immune mechanisms. Gut microbiome metabolites like SCFAs also affect serotonin signaling by regulating 5-HT receptor expression by increasing the mRNA expression of 5-HTR1A, 2B, and 5-HT7. Bacterial extracellular vesicles (EVs) that are hypothesized to permeate the blood–brain barrier (BBB) cause an increase in colonic and hippocampal serotonin levels [71].

4. Influence of microbiome on brain function and behavior

Bacteria within the gut microbiome play crucial roles in the maintenance of gut epithelium integrity, digestion, metabolism, synthesis of beneficial substances including vitamins, combating infection and inflammation and resisting colonization by pathogenic bacteria during conditions of good health [72]. Gut microbiome dysbiosis is being implicated in a myriad of conditions like stress, obesity, and inflammation, however, elucidating the effects of the microbiome on behavior has been especially fascinating. Such research highlights the importance of the complexity and diversity of gut microbiome in both health and disease and underscores the need for further exploration.

4.1 Human personality

Investigation of microbiome composition and diversity with respect to human personality has analyzed bacterial genera linked to human behavior and has revealed that sociability is associated with higher diversity and stress and anxiety are associated with reduced diversity. These results add a new dimension to the evidence that gut microbiome can influence the central nervous system in humans with effects on behavior and stresses the ways in which modern-day living with fewer social interactions, less time spent with nature, processed diets, and oversanitized environments might be contributing to gut dysbiosis [73].

4.2 Cognition

Microbiome mediates the plasticity of cognitive traits by altering protein expression, adult hippocampal neurogenesis and performance on cognitive tasks [74].

4.3 Physical activity

Evidence suggests that aerobic exercise improves the diversity and abundance of genera from the *Firmicutes* phylum, which may be the link between the positive effects of exercise on the gut and brain [75].

4.4 Autism spectrum disorder (ASD)

Numerous studies, stimulated by frequent gastrointestinal complaints and immune dysregulation in ASD, have suggested microbial dysbiosis in clinical populations of ASD. Increased blood levels of lipopolysaccharides with a corresponding increase in peripheral IL-6 levels have been found in ASD patients. Further, increased intestinal permeability as reported in ASD subjects and their first-degree relatives is posited to be a pathogenetic factor rather than a consequence of autistic behavior. Probiotic supplementation, oral vancomycin treatment and a modified fecal microbiome transfer have been demonstrated to have therapeutic potential in children with ASD [57].

4.5 Major depressive disorder

Serotonin in CNS is synthesized from tryptophan transported from blood and hence, tryptophan availability is critical for serotonin synthesis in brain [76]. Normal gut microbiota buffer extreme fluctuations in serotonin levels by making the serotonergic system less sensitive to variations in its precursor. Gut serotonin, which is under the control of gut microbiota, through neural routes communicates to the brainstem neurons by stimulating serotonin receptors at the terminals of vagal afferents [77]. It has been shown that vagotomy abolishes the antidepressant effects of SSRIs thus implicating the role of peripheral serotonin and vagus nerve stimulation in the regulation of depressive behavior [78] and leading to the postulation that gut microbiome through stimulation of the vagus nerve influences depressive behavior [79]. A few case–control studies have reported differences in the gut microbiome between depressed patients, and healthy controls and preliminary evidence suggests that probiotics might prove to be beneficial in patients with major depressive disorder and in healthy populations as well [80].

4.6 Bipolar disorder

Extensive research has suggested an abnormal inflammatory response in bipolar disorder and evidence is emerging for alterations in gut microbial composition of patients with bipolar disorder suggesting that gut microbial dysbiosis contributes to disease progression and cognitive impairment in bipolar disorder [81]. Gut microbial profiling in bipolar disorder patients revealed some correlation between certain genera and sleep and stress in these patients [82].

4.7 Schizophrenia

High rates of comorbidity reported in schizophrenia with autoimmune and gastrointestinal conditions, systemic low-level inflammation, and increased intestinal permeability suggest the involvement of gut microbiome. Studies that need further investigation, have identified reduced phylum *Proteobacteria* and *Gammaproteobacteria* as class-level biomarkers of schizophrenia [57]. Antipsychotic medications have been shown to alter gut microbiome and this alteration is considered to play a crucial role in adverse effects like metabolic syndrome resulting from antipsychotic medication use [83]. A positive correlation was seen between *Lactobacillus* bacterial group members and the severity of psychotic symptoms in a study in which 70% of subjects showed remission on antipsychotic treatment whereas only 28% subjects with “abnormal” microbiota experienced remission indicating thereby that gut microbiome may moderate treatment response in schizophrenia. The use of probiotics improved gastrointestinal disturbance in psychosis although improvement in symptoms of psychosis has not shown promising results [57].

4.8 Attention deficit hyperactivity disorder (ADHD)

Altered and reduced diversity of gut microbiome in children with ADHD has been reported in studies limited by sample size and concomitant methylphenidate intake [84, 85].

4.9 Anxiety and related disorders

Anxiety and related disorders viz., obsessive-compulsive disorder (OCD) and pediatric autoimmune neuropsychiatric disorder associated with streptococcal infection (PANDAS) are other areas wherein indirect evidence suggests that gut microbiome has a potential causative role and therapeutic potential [57].

5. Recent advances and potential therapeutic targets

Some of the key technologies used to investigate the complex interactions between the gut microbiome and the central nervous system include:

5.1 Multi-omics

Metagenomic sequencing allows researchers to analyze the genetic material of the entire microbial community in the gut, providing insights into the diversity and functional potential of the microbiome. Metatranscriptomics technology focuses on the RNA transcripts of the gut microbiome, revealing which genes are actively expressed and providing information on microbial functions. Metabolomics is used to study the small molecules produced by gut microbes, such as short-chain fatty acids and neurotransmitters, which can influence brain function. Multi-omics integration combining data from genomics, transcriptomics, and metabolomics can provide a comprehensive understanding of the microbiome–gut–brain axis. In microbiota profiling, 16S rRNA sequencing and shotgun metagenomic sequencing are used to identify and characterize the composition of microbial communities in the gut [86].

5.2 Functional neuroimaging

Techniques like fMRI (functional magnetic resonance imaging) can be employed to observe changes in brain activity in response to gut microbiome alterations.

5.3 Animal models

Animal studies, including germ-free and gnotobiotic models, are used to investigate the effects of specific microbial communities on behavior and brain function.

5.4 Germ-free studies

Germ-free animals are the "microbiota free" control group for the animals whose gut is conventionally colonized. They are maintained in gnobiotic units which are sterile, eliminating the chances of postnatal colonization of their GI tracts [87]. Germ-free animals are studied for social, stereotypical, and anxiety-like behaviors on exposure to novel and aversive environments (elevated plus maze, light/dark box, open field), and non-spatial and working memory tasks (novel object recognition and spontaneous alternation assessed in the T-maze) in labs. Germ-free mice have been shown to have lower levels of molecular targets like the N-methyl-d-aspartate receptors (NMDARs) in the hippocampus, or amygdala and decreased levels of brain-derived neurotrophic factor (BDNF). Apart from other findings, the results have been seen to be dependent on the time of colonization, be it in adolescence or adulthood, positing that there is a critical time period that is neurodevelopmentally sensitive to dysbiosis [88].

5.5 Antibiotics

Both *in-vitro* and *in-vivo* experiments [89] have documented the perturbation of gut microbiota with antibiotic treatment that leads to an increase in sensitivity to visceral pain, increase in gut motility and altered BDNF levels in the brain.

5.6 Probiotics and prebiotics

Probiotics are defined as live organisms which when administered in adequate amounts confer a health benefit on the host. Prebiotics are non-digestible food ingredients that selectively stimulate the growth of Lactobacilli and Bifidobacteria in the gut, hence indirectly affecting the brain function. The two main genera which are used as probiotics are Lactobacillus and Bifidobacterium. Several pre- and probiotic studies have demonstrated their beneficial effects on behavior of the host and offer novel therapeutic potential for treating mood and anxiety disorders [90]. Caution is warranted when translating and generalizing the evidence that certain pre- and/or probiotic strains are able to modulate brain function and behavior.

5.7 Microbiome manipulation

Researchers can manipulate the gut microbiome using techniques such as fecal microbiota transplantation (FMT) to assess its impact on brain health and behavior [91].

5.8 Brain-gut communication assays

These assays involve measuring biomarkers like cytokines and neuropeptides to understand how the gut and brain communicate. Finally, researchers may employ psychological tests and behavioral observations to assess the impact of the gut microbiome on mood, cognition, and other brain-related functions [92].

5.9 Potential therapeutic strategies targeting the microbiome gut-brain axis

Efforts are being put to target the vast ecosystem of the gut microbiome for a role in neuropsychiatric disorders.

5.9.1 Psychobiotic

Dinan et al. coined the term "psychobiotic" and defined it as "live organism that, when ingested in adequate amounts, produces a health benefit in patients suffering from psychiatric illness." This definition has been expanded since then to include "any exogenous influence whose effect on the brain is bacterially-mediated." Thus, psychobiotics include a range of substances that have the potential to affect microbiota–gut–brain axis signaling, including probiotics, prebiotics, symbiotics, and postbiotics. These substances can be delivered through supplements, functional foods, and improvements to dietary intake. Some microbial therapeutics have been engineered to sense a range of biomarkers and respond accordingly and are currently in clinical trials for the treatment of diabetes, inflammation, and cancers [71].

5.9.2 Probiotics

There is preliminary evidence from human studies wherein Probiotics have demonstrated their efficacy in ameliorating anxiety and depression states [93]. These findings have been supported by systematic reviews of therapeutic potential of probiotics in MDD [94].

5.9.3 Prebiotics

Prebiotics consist of fibers such as resistant starch, fructo-oligosaccharides, galacto-oligosaccharides (GOSs), and inulin which are unabsorbed in the small intestine and are selectively fermented by gut microbes. Prebiotics are also found in human milk. One study which was done in a cohort of patients suffering from IBS demonstrated a significant decrease in anxiety scores with prebiotic administration [95].

5.9.4 Synbiotics

Synbiotics are a combination of both pre- and probiotics, whereby the prebiotics improves the viability of the probiotic, providing a source of fermentable fiber as well as acting as a general prebiotic. In a recent study, a symbiotic comprising galacto-oligosaccharides (GOS) and a dual-strain probiotic (*Lactobacillus helveticus* and *B. longum*) was successfully shown to decrease scores on depression scale and positively impacted tryptophan signaling in mild to moderate MDD [96].

5.9.5 Postbiotics

Postbiotics are nonviable entities that are byproducts of bacterial fermentation and include bioactive metabolites such as SCFAs. The use of gut peptides directly as an intervention in gut–brain axis may not be feasible due to technical issues, however, targeting specific microbiota in order to modulate specific gut peptides may be a useful psychobiotic therapy. Para probiotics, or nonviable probiotics, e.g., heat-killed probiotics, can also be included in the category of postbiotics in that they contain structural components that may exert biological activity in the host. In preclinical studies, several heat-killed probiotics have described antidepressant and anxiolytic effects, with heat-killed *Lactobacillus paracasei* [97].

5.9.6 Fermented foods and diet

Fermented foods contain probiotics, prebiotics, and bacterially derived bioactives. Two of the most common strains used in the fermentation process include *Lactobacillus delbrueckii* subsp. bulgaricus and *Streptococcus thermophilus*. In dairy products, lactic acid-producing *Bifido-bacterium* and *Lactobacillus* are commonly used. Studies using fermented food interventions in humans are limited, however, there is some evidence showing ameliorations in anxiety and mood scores. Fermented milk drinks have been found to result in positive benefits in emotional processing. Fermented milk containing *Lactobacillus casei* strain Shirota prevents the onset of physical symptoms in medical students under academic stress by modulating the gut-brain interaction [98]. Mediterranean diets have well-known mental health benefits. One large-cohort, cross-sectional study in women found healthier dietary patterns to be associated with better general health scores and decreased incidence of anxiety and depression outcomes [99].

6. Ethical considerations and challenges

As with any medical intervention, ethical concerns arise regarding the use of microbiome-based interventions for mental health and neurology. Balancing potential benefits with safety and long-term effects is crucial. It is important to note that while there is promising research in this area, it is still relatively new and complex. More studies are needed to fully understand the mechanisms at play and to establish the effectiveness and safety of microbiome-based interventions for mental health and neurology.

Ethical considerations in microbiome-based treatments include issues related to informed consent, privacy and data security, equitable access, potential conflicts of interest, and the long-term effects of manipulating the microbiome. These treatments involve complex interactions with individual health and the broader ecosystem, requiring careful consideration of both short-term benefits and potential unintended consequences. One of the challenges in this field is the identification of neuroactive compounds originating from the host rather than gut microbiome due to complex communications between these two. Many of the dietary benefits on the microbiome and brain health have been attributed to anti-inflammatory effects mediated by the microbial metabolites of dietary fiber and polyphenols. Overall, it is clear that although animal studies have shown much promise, more progress is necessary before these findings can be translated for diagnostic and therapeutic benefit in patient populations.

Author details

Mushtaq Margoob[1,2*], Shazia Kouser[2] and Neelofer Jan[2]

1 Advanced Institute for Management of Stress and Lifestyle-related Problems (AIMS), Srinagar, Kashmir, India

2 Department of Psychiatry, Institute of Mental Health and Neurosciences-Kashmir, Government Medical College, Srinagar, Kashmir, India

*Address all correspondence to: mushtaqmargoob2@gmail.com

References

[1] Katzung BG. Basic and Clinical Pharmacology. McGraw Hill; 2012

[2] Rapport MM, Green AA, Page IH. Serum vasoconstrictor (serotonin): iii. Chemical inactivation. Journal of Biological Chemistry. 1948;**176**(3):1237-1241

[3] Hornung J-P. The human raphe nuclei and the serotonergic system. Journal of Chemical Neuroanatomy. 2003;**26**(4):331-343

[4] Schatzberg AF, Nemeroff CB. The American Psychiatric Association Publishing Textbook of Psychopharmacology. American Psychiatric Pub; 2017

[5] Tao Z-Y et al. The role of descending pain modulation in chronic primary pain: Potential application of drugs targeting serotonergic system. Neural Plasticity. 2019;**2019**:1389296

[6] Krautkramer KA, Fan J, Bäckhed F. Gut microbial metabolites as multi-kingdom intermediates. Nature Reviews Microbiology. 2021;**19**(2):77-94

[7] Deniker P et al. Measurement of the plasmatic level of free and protein-bound tryptophan in mental pathology. Neuropsychobiology. 1980;**6**(3):132-139

[8] Höglund E, Øverli Ø, Winberg S. Tryptophan metabolic pathways and brain serotonergic activity: A comparative review. Frontiers in Endocrinology. 2019;**10**:158

[9] Le Floch N, Otten W, Merlot E. Tryptophan metabolism, from nutrition to potential therapeutic applications. Amino Acids. 2011;**41**:1195-1205

[10] Zhu X et al. Comprehensive bibliometric analysis of the kynurenine pathway in mood disorders: Focus on gut microbiota research. Frontiers in Pharmacology. 2021;**12**:687757

[11] Miri S et al. Neuromicrobiology, an emerging neurometabolic facet of the gut microbiome? Frontiers in Microbiology. 2023;**14**:1098412

[12] Richard DM et al. L-tryptophan: Basic metabolic functions, behavioral research and therapeutic indications. International Journal of Tryptophan Research. 2009;**2**:S2129

[13] Davis I, Liu A. What is the tryptophan kynurenine pathway and why is it important to neurotherapeutics? Expert Review of Neurotherapeutics. 2015;**15**(7):719-721

[14] Martin CA, Krantz DE. Drosophila melanogaster as a genetic model system to study neurotransmitter transporters. Neurochemistry International. 2014;**73**:71-88

[15] Wray NH, Rasenick MM. Lipid rafts in psychiatry. Advances in Pharmacology. 2019;**86**:21-45

[16] Schildkraut JJ. The catecholamine hypothesis of affective disorders: A review of supporting evidence. American Journal of Psychiatry. 1965;**122**(5):509-522

[17] Mulinari S. Monoamine theories of depression: Historical impact on biomedical research. Journal of the History of the Neurosciences. 2012;**21**(4):366-392

[18] Cassano G, Marazziti D. Is depression a disorder of a receptor superfamily?

A critical review of the receptor theory of depression and the appraisal of a new heuristic model. European Psychiatry. 1992;7(6):259-270

[19] Moncrieff J et al. The serotonin theory of depression: A systematic umbrella review of the evidence. Molecular Psychiatry. 2022;**28**:1-14

[20] Berlucchi G, Buchtel HA. Neuronal plasticity: Historical roots and evolution of meaning. Experimental Brain Research. 2009;**192**:307-319

[21] Ming G-L, Song H. Adult neurogenesis in the mammalian brain: Significant answers and significant questions. Neuron. 2011;**70**(4):687-702

[22] Kumar A et al. Adult neurogenesis in humans: A review of basic concepts, history, current research, and clinical implications. Innovations in Clinical Neuroscience. 2019;**16**(5-6):30

[23] Duman RS, Li N. A neurotrophic hypothesis of depression: Role of synaptogenesis in the actions of NMDA receptor antagonists. Philosophical Transactions of the Royal Society B: Biological Sciences. 2012;**367**(1601):2475-2484

[24] Kraus C et al. Serotonin and neuroplasticity–links between molecular, functional and structural pathophysiology in depression. Neuroscience & Biobehavioral Reviews. 2017;**77**:317-326

[25] Maddaloni G et al. Serotonin depletion causes valproate-responsive manic-like condition and increased hippocampal neuroplasticity that are reversed by stress. Scientific Reports. 2018;**8**(1):11847

[26] Wilson CC, Faber KM, Haring JH. Serotonin regulates synaptic connections in the dentate molecular layer of adult rats via 5-HT1a receptors: Evidence for a glial mechanism. Brain Research. 1998;**782**(1-2):235-239

[27] Liu B et al. From serotonin to neuroplasticity: Evolvement of theories for major depressive disorder. Frontiers in Cellular Neuroscience. 2017;**11**:305

[28] Gandhi AB et al. Neuroplasticity improves bipolar disorder: A review. Cureus. 2020;**12**(10):e11241

[29] Meneses A. 5-HT systems: Emergent targets for memory formation and memory alterations. Reviews in the Neurosciences. 2013;**24**(6):629-664

[30] Kucwaj-Brysz K et al. Chemical update on the potential for serotonin 5-HT6 and 5-HT7 receptor agents in the treatment of Alzheimer's disease. Bioorganic & Medicinal Chemistry Letters. 2021;**49**:128275

[31] Ramírez MJ. 5-HT 6 receptors and Alzheimer's disease. Alzheimer's Research & Therapy. 2013;**5**:1-8

[32] Blundell J. Serotonin and appetite. Neuropharmacology. 1984;**23**(12):1537-1551

[33] Gershon M et al. Serotonergic neurons in the peripheral nervous system: Identification in gut by immunohistochemical localization of tryptophan hydroxylase. Proceedings of the National Academy of Sciences. 1977;**74**(7):3086-3089

[34] Yadav VK et al. Leptin regulation of bone mass, appetite and energy expenditure relies on its ability to inhibit serotonin synthesis in the brainstem. Cell. 2009;**138**(5):976

[35] Fourtillan JB. Role of melatonin in the induction and maintenance of sleep.

Dialogues in Clinical Neuroscience. 2022;**4**(4):395-401

[36] España RA, Scammell TE. Sleep neurobiology from a clinical perspective. Sleep. 2011;**34**(7):845-858

[37] Hiller W, Fichter MM, Rief W. A controlled treatment study of somatoform disorders including analysis of healthcare utilization and cost-effectiveness. Journal of Psychosomatic Research. 2003;**54**(4):369-380

[38] Häuser W et al. The role of antidepressants in the management of fibromyalgia syndrome: A systematic review and meta-analysis. CNS Drugs. 2012;**26**:297-307

[39] Dorszewska J et al. Serotonin in neurological diseases. In: Kaneez FS, editor. Serotonin—A Chemical Messenger between all Types of Living Cells. Books on Demand; 2017. pp. 219-239

[40] Petrucci AN et al. Serotonin and sudden unexpected death in epilepsy. Experimental Neurology. 2020;**325**:113145

[41] Theodore WH. Does serotonin play a role in epilepsy? Epilepsy Currents. 2003;**3**(5):173-177

[42] Politis M, Niccolini F. Serotonin in Parkinson's disease. Behavioural Brain Research. 2015;**277**:136-145

[43] Fox SH, Chuang R, Brotchie JM. Serotonin and Parkinson's disease: On movement, mood, and madness. Movement Disorders. 2009;**24**(9):1255-1266

[44] Banskota S, Ghia J-E, Khan WI. Serotonin in the gut: Blessing or a curse. Biochimie. 2019;**161**:56-64

[45] Jones LA et al. The ever-changing roles of serotonin. The International Journal of Biochemistry & Cell Biology. 2020;**125**:105776

[46] Singh P, Dutta SR, Guha D. Gastric mucosal protection by aegle marmelos against gastric mucosal damage: Role of enterochromaffin cell and serotonin. Saudi Journal of Gastroenterology. 2015;**21**(1):35

[47] Ahern GP. 5-HT and the immune system. Current Opinion in Pharmacology. 2011;**11**(1):29-33

[48] Shajib MS et al. Characterization of serotonin signaling components in patients with inflammatory bowel disease. Journal of the Canadian Association of Gastroenterology. 2019;**2**(3):132-140

[49] Zamani A, Qu Z. Serotonin activates angiogenic phosphorylation signaling in human endothelial cells. FEBS Letters. 2012;**586**(16):2360-2365

[50] Rinninella E et al. What is the healthy gut microbiota composition? A changing ecosystem across age, environment, diet, and diseases. Microorganisms. 2019;**7**(1):14

[51] Bäckhed F et al. Dynamics and stabilization of the human gut microbiome during the first year of life. Cell Host & Microbe. 2015;**17**(5):690-703

[52] Turnbaugh PJ et al. A core gut microbiome in obese and lean twins. Nature. 2009;**457**(7228):480-484

[53] Singhal M et al. Serotonin transporter deficiency is associated with dysbiosis and changes in metabolic function of the mouse intestinal microbiome Scientific Reports. 2019;**9**(1):2138

[54] Shao Y et al. Stunted microbiota and opportunistic pathogen colonization in caesarean-section birth. Nature. 2019;**574**(7776):117-121

[55] Robertson RC et al. The human microbiome and child growth–first 1000 days and beyond. Trends in Microbiology. 2019;**27**(2):131-147

[56] Langdon A, Crook N, Dantas G. The effects of antibiotics on the microbiome throughout development and alternative approaches for therapeutic modulation. Genome Medicine. 2016;**8**(1):1-16

[57] Van Ameringen M et al. The gut microbiome in psychiatry: A primer for clinicians. Depression and Anxiety. 2019;**36**(11):1004-1025

[58] Bastiaanssen TF et al. Volatility as a concept to understand the impact of stress on the microbiome. Psychoneuroendocrinology. 2021;**124**:105047

[59] Bailey MT, Coe CL. Maternal separation disrupts the integrity of the intestinal microflora in infant rhesus monkeys. Developmental Psychobiology: The Journal of the International Society for Developmental Psychobiology. 1999;**35**(2):146-155

[60] Zijlmans MA et al. Maternal prenatal stress is associated with the infant intestinal microbiota. Psychoneuroendocrinology. 2015;**53**:233-245

[61] Clarke SF et al. Exercise and associated dietary extremes impact on gut microbial diversity. Gut. 2014;**63**(12):1913-1920

[62] Bressa C et al. Differences in gut microbiota profile between women with active lifestyle and sedentary women. PLoS One. 2017;**12**(2):e0171352

[63] Ghosh TS, Shanahan F, O'Toole PW. The gut microbiome as a modulator of healthy ageing. Nature Reviews Gastroenterology & Hepatology. 2022;**19**(9):565-584

[64] Mayer EA, Nance K, Chen S. The gut–brain axis. Annual Review of Medicine. 2022;**73**:439-453

[65] Fülling C, Dinan TG, Cryan JF. Gut microbe to brain signaling: What happens in vagus.... Neuron. 2019;**101**(6):998-1002

[66] Appleton J. The gut-brain axis: Influence of microbiota on mood and mental health. Integrative Medicine: A Clinician's Journal. 2018;**17**(4):28

[67] Raber J et al. Corticotropin-releasing factor and adrenocorticotrophic hormone as potential central mediators of OB effects. Journal of Biological Chemistry. 1997;**272**(24):15057-15060

[68] Picciotto M. Galanin–25 years with a multitalented neuropeptide: Galanin and addiction. Cellular and Molecular Life Sciences. 2008;**65**:1872-1879

[69] Holzer P, Reichmann F, Farzi A. Neuropeptide Y, peptide YY and pancreatic polypeptide in the gut–brain axis. Neuropeptides. 2012;**46**(6):261-274

[70] Mayer EA, Tillisch K, Gupta A. Gut/brain axis and the microbiota. The Journal of Clinical Investigation. 2015;**125**(3):926-938

[71] Everett BA, Tran P, Prindle A. Toward manipulating serotonin signaling via the microbiota–gut–brain axis. Current Opinion in Biotechnology. 2022;**78**:102826

[72] Bidell MR, Hobbs AL, Lodise TP. Gut microbiome health and dysbiosis: A clinical primer. Pharmacotherapy: The

Journal of Human Pharmacology and Drug Therapy. 2022;**42**(11):849-857

[73] Johnson KV-A. Gut microbiome composition and diversity are related to human personality traits. Human Microbiome Journal. 2020;**15**:100069

[74] Davidson GL et al. The gut microbiome as a driver of individual variation in cognition and functional behaviour. Philosophical Transactions of the Royal Society B: Biological Sciences. 2018;**373**(1756):20170286

[75] Dalton A, Mermier C, Zuhl M. Exercise influence on the microbiome–gut–brain axis. Gut Microbes. 2019;**10**(5):555-568

[76] Lukić I et al. Role of tryptophan in microbiota-induced depressive-like behavior: Evidence from tryptophan depletion study. Frontiers in Behavioral Neuroscience. 2019;**13**:123

[77] Browning KN. Role of central vagal 5-HT3 receptors in gastrointestinal physiology and pathophysiology. Frontiers in Neuroscience. 2015;**9**:413

[78] McVey Neufeld K-A et al. Oral selective serotonin reuptake inhibitors activate vagus nerve dependent gut-brain signalling. Scientific Reports. 2019;**9**(1):14290

[79] Yano JM et al. Indigenous bacteria from the gut microbiota regulate host serotonin biosynthesis. Cell. 2015;**161**(2):264-276

[80] Akkasheh G et al. Clinical and metabolic response to probiotic administration in patients with major depressive disorder: A randomized, double-blind, placebo-controlled trial. Nutrition. 2016;**32**(3):315-320

[81] Dai W et al. Gut microbial dysbiosis and cognitive impairment in bipolar disorder: Current evidence. Frontiers in Pharmacology. 2022;**13**:893567

[82] Long-Smith C et al. Microbiota-gut-brain axis: New therapeutic opportunities. Annual Review of Pharmacology and Toxicology. 2020;**60**:477-502

[83] Davey K et al. Antipsychotics and the gut microbiome: Olanzapine-induced metabolic dysfunction is attenuated by antibiotic administration in the rat. Translational Psychiatry. 2013;**3**(10):e309-e309

[84] Aarts E et al. Gut microbiome in ADHD and its relation to neural reward anticipation. PLoS One. 2017;**12**(9):e0183509

[85] Prehn-Kristensen A et al. Reduced microbiome alpha diversity in young patients with ADHD. PLoS One. 2018;**13**(7):e0200728

[86] Daliri EB-M et al. Challenges and perspective in integrated multi-omics in gut microbiota studies. Biomolecules. 2021;**11**(2):300

[87] Neufeld K et al. Reduced anxiety-like behavior and central neurochemical change in germ-free mice. Neurogastroenterology & Motility. 2011;**23**(3):255-e119

[88] Diamond B et al. It takes guts to grow a brain: Increasing evidence of the important role of the intestinal microflora in neuro-and immune-modulatory functions during development and adulthood. BioEssays. 2011;**33**(8):588-591

[89] Bercik P et al. The intestinal microbiota affect central levels of brain-derived neurotropic factor and behavior in mice. Gastroenterology. 2011;**141**(2):599-609. e3

[90] Dinan TG, Stanton C, Cryan JF. Psychobiotics: A novel class of psychotropic. Biological Psychiatry. 2013;**74**(10):720-726

[91] Li N et al. Fecal microbiota transplantation relieves gastrointestinal and autism symptoms by improving the gut microbiota in an open-label study. Frontiers in Cellular and Infection Microbiology. 2021;**11**:948

[92] Zhang B et al. Brain–gut communications via distinct neuroendocrine signals bidirectionally regulate longevity in *C. elegans*. Genes & Development. 2018;**32**(3-4):258-270

[93] Benton D, Williams C, Brown A. Impact of consuming a milk drink containing a probiotic on mood and cognition. European Journal of Clinical Nutrition. 2007;**61**(3):355-361

[94] Yong SJ et al. Antidepressive mechanisms of probiotics and their therapeutic potential. Frontiers in Neuroscience. 2020;**13**:1361

[95] Azpiroz F et al. Effects of sc FOS on the composition of fecal microbiota and anxiety in patients with irritable bowel syndrome: A randomized, double blind, placebo controlled study. Neurogastroenterology & Motility. 2017;**29**(2):e12911

[96] Majeed M et al. Bacillus coagulans MTCC 5856 for the management of major depression with irritable bowel syndrome: A randomised, double-blind, placebo controlled, multi-centre, pilot clinical study. Food & Nutrition Research. 2018:62

[97] Kavvadia M et al. Psychobiotics as integrative therapy for neuropsychiatric disorders with special emphasis on the microbiota-gut-brain axis. Biomedicine and Prevention. 2017;**2**(8)

[98] Takada M et al. Probiotic *Lactobacillus casei* strain shirota relieves stress-associated symptoms by modulating the gut–brain interaction in human and animal models. Neurogastroenterology & Motility. 2016;**28**(7):1027-1036

[99] Merra G et al. Influence of mediterranean diet on human gut microbiota. Nutrients. 2020;**13**(1):7

Chapter 3

The Platelet Serotonergic System and the Search for New Biomarkers and Therapeutic Options for Diverse Diseases

Alfredo Briones-Aranda, Josselin Corzo-Gómez, Diana Casique-Aguirre and Mauricio Megchún-Hernández

Abstract

The latest advances in basic and clinical research on the main components of the platelet serotonergic system are presently reviewed. These components consist of serotonin (5-HT), enzymes that participate in 5-HT metabolism, the serotonin transporter (SERT), and 5-HT1A, 5-HT2A, 5-HT3, and 5-HT4 receptors (each with their corresponding mechanism of intracellular transduction). An additional focus is on related biomarkers or drugs for the diagnosis or treatment of the pathophysiology of diverse disorders such as depression, anxiety, hemorrhagic dengue, coagulopathy generated by COVID-19, myocardial infarction, and preeclampsia. The drugs analyzed include serotonin reuptake inhibitors and serotonergic drugs that act on 5-HT receptors. Through the platelet serotonergic system, serotonergic drugs not only interact with the central nervous system but also may participate in coagulation, vascular permeability, and peripheral vascular resistance, which has many implications. Finally, perspectives are offered for future research on biomarkers and new therapeutic targets.

Keywords: platelets, serotonin, pathophysiology, serotonin transporter, 5-HT1A, 5-HT2A

1. Introduction

Scientific advances in biomedical research have allowed for the discovery of numerous factors in the human organism related to ligand-receptor interactions and intracellular signaling mechanisms. The new insights have provided a deeper comprehension of the interrelationship of diverse complex physiological systems on a molecular and systemic level. For example, recent studies on the platelet serotonergic system have created new opportunities based on a better understanding the pathophysiology of severe dengue, coagulopathy associated with COVID-19, myocardial infarction, preeclampsia, and other diseases.

IntechOpen

Platelets are anucleate blood cells with a biconvex discoid structure and a diameter of 2–4 μm. They develop from megakaryocytes of the bone marrow and are released into the bloodstream through projections of pseudopodia known as proplatelets [1]. Each megakaryocyte is able to produce ~1000 to 3000 platelets as a result of multiple divisions [2]. Platelets contain 5–8 mitochondria that synthesize various proteins involved in platelet activation and apoptosis [3]. Unactivated platelet cells have an average circulation lifespan of 8–10 days before being eliminated by the spleen. People generally have 150–400 × 10^9 platelets per liter of peripheral blood [1].

A key molecule of the platelet serotonergic system is serotonin (5-HT), which belongs to the family of indolamines. It is composed of an indole ring hydroxylated at position 5 and an ethylamine lateral chain [4]. Apart from 5-HT, other important components of the platelet serotonergic system are the serotonin transporter (SERT) responsible for the recapture of 5-HT, a broad range of serotonin membrane receptors, and enzymes that participate in 5-HT metabolism [5]. The elements of the platelet serotonergic system are located in the central nervous system (CNS) [6] enterochromaffin cells, and diverse tissues (e.g., pancreatic, pulmonary, hepatic, and hematopoietic) [7–9]. This composition of the serotonergic system regulates most of the physiological functions of the organism that have a role in the maintenance of homeostasis.

At the beginning of the 20th century, platelets were thought to be the only source of 5-HT in mammals. About 100 years ago, Dr. Vittorio Erspamer was looking for substances capable of causing the contraction of smooth muscle. In 1930, he identified such a compound, which he called enteramine, in the gastric mucosa of rabbits. Almost two decades later (in 1948), a vasoconstrictive substance, denominated serotonin, was isolated from human serum in the laboratory of Dr. Irving Page. The structure of serotonin was established as 5-HT in 1949, and in 1952 it was confirmed that enteramine and serotonin are the same compound [7].

Regarding the biosynthesis of 5-HT, the first step is the conversion of L-tryptophan (L-Trp) to 5-hydroxytryptophan (5-HTP), catalyzed by the rate-limiting enzyme tryptophan hydroxylase. Subsequently, 5-HTP is transformed into 5-HT by another enzyme, aromatic L-amino acid decarboxylase (4). Additionally, L-Trp can be metabolized in the kynurenine pathway to nicotinamide, a fundamental component in the formation of two molecules involved in important redox reactions in cell metabolism: nicotinamide adenine dinucleotide (NAD^+) and nicotinamide adenine dinucleotide phosphate ($NADP^+$). The majority of 5-HT in the human organism is produced and stored in the gastrointestinal tract (**Figure 2**). From this location, it enters the bloodstream and by means of SERT is rapidly internalized into platelets [5, 10] (5), then stored in the dense granules of these cells. Only a small fraction of 5-HT (<1%) remains free in plasma [11].

The 5-HT secreted into plasma by platelets binds to many different membrane receptors, which are classified into seven families designated as 5-HT1–7. There are distinct subtypes for each family, denoted by adding a letter to the name (e.g., 5-HT1A, or 5-HT receptor type 1, subtype A) [12, 13]. Among the subtypes of 5-HT receptors (5-HTRs) on the platelet membrane, those most commonly studied are 5-HT1A [14, 15], 5-HT2A [16], 5-HT3 [17] and 5-HT4 [18]. They will hereafter be discussed along with their ligands and mechanisms of signal transduction.

Platelets participate in the defense against pathogens [19] as well as in the exacerbation of autoimmune diseases [20, 21]. For example, in the presence of certain infectious agents or inflammatory stimuli, activated platelets trigger the innate and adaptive immune response by releasing granules containing 5-HT, β-defensin [22, 23] prostaglandins, and other inflammatory mediators. 5-HT released into plasma stimulates receptors with the capacity to promote vasoconstriction and therefore increase peripheral vascular resistance [24], which is characteristic of a broad range of hypertensive pathologies [25, 26].

In summary, platelets may be regarded as circulating neurons with a complex communication network that enables them to interact with cells in many types of tissues to regulate diverse physiological processes. Hence the platelet serotonergic system acts as an axis of autocrine and paracrine communication [27] capable of mediating inflammation, blood pressure, and the immune response.

These insights into the platelet serotonergic system imply advantages, disadvantages, and challenges in relation to medical applications. Considering the platelet serotonergic system as the common denominator for the CNS and diverse peripheral tissues (e.g., in the gastrointestinal tract, the hematopoietic system, and the cardiovascular apparatus), platelets can be viewed as "circulating mirrors" of neurons and innate immune cells [28, 29]. That is, the activity of the platelet serotonergic system appears to be a reflection of what occurs in peripheral tissues during various pathological processes: neuropsychological disorders (e.g., stress, depression, anxiety, and epilepsy), coagulopathies associated with infectious diseases [e.g., dengue [30] and COVID-19 [31], and hypertensive disorders such as preeclampsia [24] and acute myocardial infarction [32]. For instance, the aforementioned neuropsychological disorders involve increased permeability of the blood-brain barrier, which gives rise to a greater filtration of platelets towards the CNS. In the brain, platelets release 5-HT and other chemical mediators, exacerbating neural electrical activity. In cases of epilepsy, this release of 5-HT could increase the severity of convulsions [33]. Thus, compounds capable of slowing the course of platelet activation might be useful for treating epilepsy. A deeper understanding of the platelet serotonergic system should allow for the development of new biomarkers for diagnosis and new drugs (e.g., agonists or antagonists of 5-HTRs) with enhanced therapeutic activity.

Regarding the disadvantages of the close relation between the role of the platelet serotonergic system in the CNS and peripheral tissues, drugs targeting the brain to treat anxiety, depression, and epilepsy might also act on the periphery and cause adverse effects. For example, platelet function can be affected by serotonin reuptake inhibitors (SRIs) such as fluoxetine [34] and valproic acid. The latter is an agonist of 5-HT2A used to treat epilepsy [35] and may under certain conditions produce thrombocytopenia [36].

A challenge for future research is the design and development of drugs able to interact selectively with the serotonergic system in the CNS or peripheral tissues. Moreover, it is crucial to gain insights into the dynamics of platelet activity leading to the release of 5-HT into plasma in order to better comprehend the kinetics of 5-HT in the pathophysiology of severe dengue, high blood pressure, preeclampsia, and acute myocardial infarction.

2. Transport and receptors of 5-HT in platelets

SERT, a member of the family of transporters of solutes dependent on Na+/Cl_6 (the solute carrier 6, or SLC6), controls intra- and extracellular concentrations of 5-HT [37, 38]. Its main function is to modulate peripheral signaling and homeostasis [39]. Given that 5-HT is a protonated molecule under physiological conditions, it is incapable of crossing the lipidic bilayer of cell membranes. Thus, the proper balance of its absorption into platelets and release from the same depends on the regulation of bidirectional transport carried out by SERT [37]. As the plasma level of 5-HT increases, SERT expression is upregulated on the platelet membrane and vice versa. Hence, regulating SERT expression is a plausible strategy for controlling plasma 5-HT levels [40].

In contrast, SRIs (e.g., fluoxetine and sertraline) prescribed to patients diagnosed with severe depression have been tested in animal models to assess their possible adverse effects. The principal clinical and biochemical changes are related to hemostasis, which is linked to the concentration of 5-HT in platelets. The use of SRIs has been reported to induce both low [41, 42] and high concentrations of 5-HT [43]. This discrepancy can be explained by the utilization of different drug schemes, assay methodologies, and/or techniques for the determination of 5-HT.

According to a study by Linder et al., veins (unlike arteries) do not have functional SERT. Nevertheless, they are capable of absorbing a significantly greater quantity of 5-HT than arteries, suggesting that their capture of 5-HT, independent of SERT [38] involves other monoamine transporters (e.g., the norepinephrine transporter and the dopamine active transporter) that are similar to SERT in their mechanism of absorption, function, and regulation [37, 44]. Further research is needed on the location and function of SERT and other monoamine transporters in the cardiovascular system in order to explore possible pharmacological targets for the treatment of vascular disorders.

The effects of 5-HT are mediated by specific 5-HTRs [45]. The 5-HT2A (type 2, subtype A) receptor is best characterized on the platelet membrane and in the cells of vascular smooth muscle tissue [46]. After binding with 5-HT2A, 5-HT is released into the bloodstream, resulting in the activation of nearby platelets and greater platelet aggregation to form thrombi during the inflammatory response. Moreover, there is evidence of a vasoconstrictor effect elicited by the activation of 5-HT2A receptors in cells of vascular smooth muscle cells, and a vasodilator effect stemming from the activation of 5-HT1B in endothelial cells through a mechanism dependent on nitric oxide (NO) [47].

Based on these findings, the antagonist sarpogrelate was tested on 5-HT2A receptors, observing a decreased serum concentration of 5-HT [48]. Such an effect could possibly diminish the formation of thrombi, thus reducing the risk of acute myocardial infarction. However, the search for a selective compound is complicated by the structural homology (46–50%) of 5-HT2A with the 5-HT2B and 5-HT2C subtypes [49]. Activation of 5-HT2B leads to the contraction of smooth muscle tissue of the stomach and muscular tissue in the human intestine. It also takes part in the relaxation of the endothelial tissue of veins [50, 51]. The 5-HT2C receptor has not been detected in the cardiovascular system and its activity is as yet unknown due to the lack of selective ligands [52].

The activation of 5-HT1A causes a drop in blood pressure and the heart rate [53], indicating a physiological role in the activation of the vagus nerve signaling the heart [54] and bladder [55]. The evaluation of antagonists of 5-HT1A has demonstrated a dose-dependent vasoconstriction, thus elevating the blood pressure and heart rate. It would then seem that selective agonists of 5-HT1A should be able to lower blood pressure, reduce the activity of the sympathetic nervous system, and increase stimulation of the vagus nerve [56].

In a similar sense, cardiopulmonary reflexes have been examined in relation to the brain and peripheral 5-HT1A receptors, which are known to modulate the recapture of 5-HT. Altered expression of 5-HT1A has been found on the platelet membrane of patients with major depressive disorder, with more severe depression associated with an overexpression of the receptor and decreased activity of 5-HT in platelets [14].

There are few reports on the function of 5-HT1B/1D subtypes. A study based on a rat model suggests that their activation may produce vasodilation and a drop in blood pressure [47, 52].

5-HT3 receptors are pentameric ion channels. They are controlled by a ligand consisting of five subunits generated by the alternative splicing of the RNA sequence of 5-HT3A-E [57]. Each subunit comprises ~450 amino acids and weighs an average of 50 kDa. These subunits are arranged around a central sodium, potassium, or calcium ion-conducting pore. In relation to other serotonin receptors, 5-HT3 has a distinct molecular structure and signaling pathway [58, 59]. Activation by 5-HT causes the ionic pore to open, and the entering current activates and subsequently desensitizes the receptor [60, 61]. 5-HT3 receptors are located in the spinal nucleus of the trigeminal nerve, the area postrema, and the solitary tract [62], areas of the CNS linked to emesis. They are also expressed on the surface of activated platelets and are associated with platelet aggregation. Hence, they likely contribute to thrombosis, although further investigation is necessary to clarify their physiological and pathophysiological role in platelet function [17, 63].

The affinity of 5-HT for 5-HT3 receptors in schizophrenic patients has been found to be four times greater than in healthy individuals. This affinity is related to the participation of isoform D of the amino acid serine (D-serine) in ligand-receptor binding [64]. Future research is needed on other possible structural and functional alterations in 5-HT3 receptors located on the platelets of schizophrenic patients in order to find new biomarkers and treatment targets. The two known therapeutic effects of targeting 5-HT3 receptors are related to nausea/emesis and irritable bowel syndrome. Regarding the former condition, ondansetron is an antagonist of 5-HT3 receptors that effectively relieves emesis induced by chemotherapy [65].

Regarding 5-HT4 receptors, the seven variants described in the literature differ in the sequence of the C-terminal segment. They have been detected at the central and peripheral level [65]. While their activation in the brain appears to influence long-term depression [66], in the intestine they can affect peristalsis, motility, and secretion [67]. 5-HT4 receptors are expressed on enterochromaffin cells in the gastrointestinal tract and on neurons of the enteric nervous system. The binding of 5-HT to these receptors facilitates the release of acetylcholine and the relaxation of the colon [68, 69], effects produced by the partial agonist tegaserod, which affects diverse physiological functions of the gastrointestinal tract [68, 70]. The expression level on platelet membranes is much lower for 5-HT4 than 5-HT2. Accordingly, there is an approximately 8-fold greater level of platelet mRNA encoding 5-HT2 than 5-HT4 receptors [18].

3. Serotonin receptor signaling pathways

The activation of 5-HTRs by 5-HT leads to the triggering of highly regulated signaling pathways capable of transmitting signals from the surrounding environment to the interior of the cell. Since signaling pathways elicit effector responses responsible for orchestrating many essential physiological processes of the human organism, they are a crucial factor in drug development. The most relevant signaling pathways activated by 5-HT are discussed hereafter, with the main focus on serotonin platelet receptors, and secondarily on those found on other types of cells.

Once released by the cells of the gastrointestinal tract, 5-HT is absorbed by SERT [71], thus allowing it to be dispersed into the bloodstream, captured by platelets, and transported to peripheral tissues. About 95% of 5-HT in the blood is stored in platelet dense granules, which also contain ATP, ADP, and Ca^{2+} [72, 73]. The various inhibitors of the capture of 5-HT by platelets include fluoxetine, sertraline, paroxetine, escitalopram, and citalopram (**Figure 1**). Although the specific mechanism of action of citalopram has not yet been defined, it is reportedly not through the inhibition of SERT.

3.1 5-HT1A receptors

5-HT1A receptors are coupled to the Gαi/o subunit of intracellular guanine nucleotide-binding proteins (G proteins), giving rise to the inhibition of adenylyl cyclase and the regulation of protein kinase A and cyclic adenosine monophosphate (PKA/cAMP) through descending pathways. Signaling triggered by these receptors increases the phosphorylation of extracellular signal-regulated kinases (ERKs) by means of a pathway dependent on phosphatidylinositol-3-kinase and serine/threonine kinase Akt (PI3K/Akt) (**Figure 1**) [74].

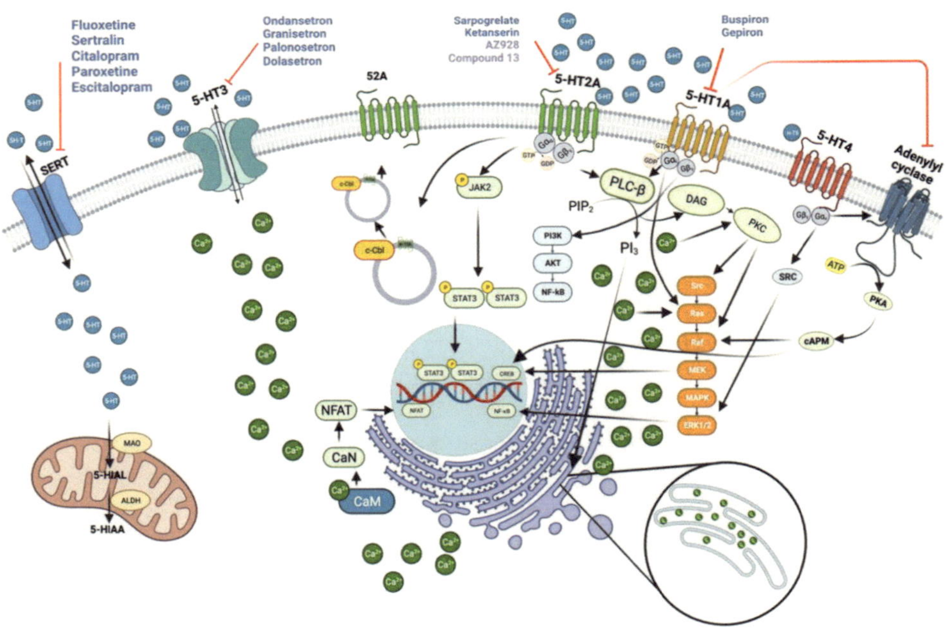

Figure 1.
The serotonin signaling pathway. Serotonin (5-HT), transported in and out of platelets by the serotonin transporter (SERT), triggers signaling by means of its receptors, including 5-HT1A, 5-HT2A, and 5-HT4 (which are G protein-coupled receptors, or GPCRs) as well as 5-HT3. The receptors illustrated on the platelet cell have significant functions in numerous cell types. Multiple signaling pathways (e.g., intracellular calcium release) are activated by these receptors. Signaling through such receptors has been associated with a variety of diseases. Drugs and compounds that interfere with the activity of 5-HT receptors have been shown to affect signaling, thus producing therapeutic and/or adverse effects. The blue circles represent 5-HT, the black arrows indicate stimulation, and the red blunt arrows portray inhibition [18, 57, 61, 74–84]. Created with BioRender.com.

3.2 5-HT2A receptors

The binding of 5-HT to 5-HT2A receptors generates conformational changes that activate the Gαq subunit by the GDP/GTP exchange reaction. The resulting GTP activates the phospholipase C-β (PLC-β) enzyme by la dissociation of complex Gβγ [75]. The PLC-β enzyme splits the phospholipid phosphatidylinositol 4,5-bisphosphate (PIP_2) into phosphatidylinositol diacylglycerol (DAG) and inositol triphosphate (IP3). DAG activates protein kinase C (PKC) on the membrane, while IP3 induces the release of Ca^{2+} from intracellular organelles (e.g., the endoplasmic reticulum). A high level of Ca^{2+} in the cytosol causes the formation of the Ca^{2+}/calmodulin complex, which activates calcineurin phosphatase (CaN). This in turn gives rise to the translocation of the nuclear factor of activated T cells (NFAT) to the nucleus, where it plays a crucial role in the CNS and in neurological diseases [75]. In response to 5-HT binding, the 5-HT2A receptor can activate another signaling pathway by means of the phosphorylization of the Janus kinase Jak2. The latter undergoes auto-phosphorylization and at the same time triggers the phosphorylization of STAT3, resulting in its translocation to the nucleus [76]. The c-Cbl carboxyl terminus protein recycles 5-HT2A receptors (**Figure 1**) [77].

The inhibition of 5-HT2A receptors is of interest to limit platelet aggregation and as a consequence prevent adverse cardiovascular events. The main 5-HT2A inhibitors in clinical use today are ketanserin and sarpogrelate. Investigation is being carried out on small molecules such as compound AZ928, derived from 6-fluorobenzo [d] isoxazole. AZ928 has demonstrated a high affinity for the receptor and antiplatelet activity superior to that of sarpogrelate, with a favorable safety profile [73]. Other compounds have a nucleus of 4-phenylcyclohexane-5-spiro- and 5-methyl-5-phenylhidantoin bound to various fragments of aryl piperazine. One example is "compound 13", reported to inhibit platelet aggregation in a manner comparable to ketanserin and with greater efficiency than sarpogrelate [79]. The antagonists of 5-HT2A are promising candidates for the future research on the development of new antiplatelet agents.

3.3 5-HT3 receptors

Pentameric complexes have a central conducting pore permeable to cations (e.g., Na^+, Ca^{2+}, or K^+) [80–82]. The activation of these receptors in platelets brings about a rapid rise in the level of cytosolic calcium, known to play a fundamental role in platelet activation and aggregation [63]. Contrarily, resveratrol reduces the release of stored Ca^{2+} ions and at the same time inhibits the entrance of Ca^{2+} into platelets, thus inhibiting platelet aggregation prompted by thrombin (**Figure 2**) [83]. Further investigation is required to clarify the mechanism of the resveratrol-induced decrease in the level of Ca^{2+} ions in order to explore the feasibility of administering the drug to generate an effect opposite to the response of the activated receptor. The antagonists of 5-HT3 receptors include ondansetron, granisetron, tropisetron, palonosetron, dolasetron, and azasetron (**Figure 1**) [57].

3.4 5-HT4 receptors

5-HT4 receptors have been identified on human platelets (as illustrated in **Figure 1** and 2) [18]. Because of being coupled to G proteins, 5-HT4 receptors promote the production of cAMP and PKA by stimulating the activation of adenylyl cyclase. These receptors are also able to trigger metabolic pathways involving the

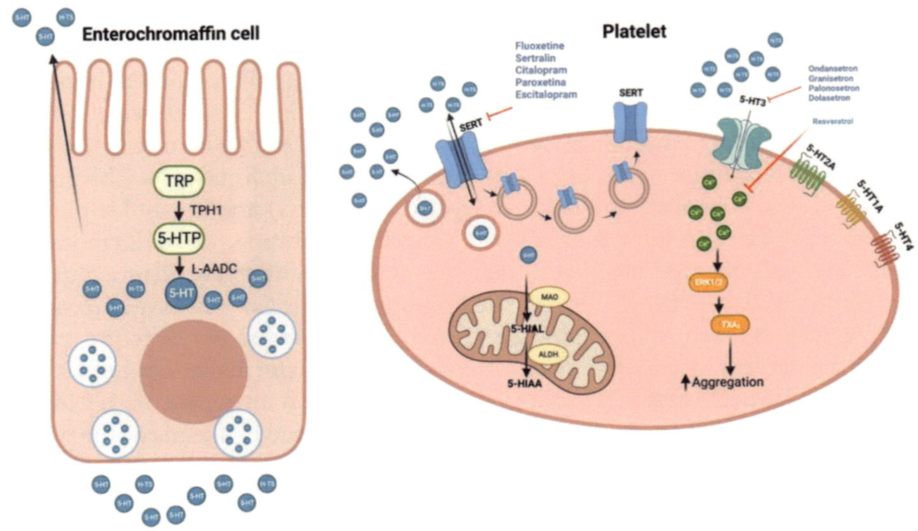

Figure 2.
The peripheral serotonergic system. Left: Synthesis of 5-HT in enterochromaffin cells, which transform tryptophan (TRP) into 5-hydroxytryptophan (5-HTP) through the enzyme tryptophan hydroxylase 1 (TPH1). Subsequently, 5-HTP is converted into 5-hydroxytryptamine (5-HT) by aromatic L-amino acid decarboxylase (L-AADC). Right: Platelets are activated by various serotonin receptors on their membrane (5-HT1A, 5-HT2A, 5-HT3, and 5-HT4). The storage of 5-HT in platelets takes place by means of the serotonin transporter (SERT), while the metabolism of 5-HT occurs through the enzyme monoamine oxidase (MAO). The latter transforms 5-HT into 5-hydroxy-indoleacetaldehyde (5-HIAL), which is oxidized by NAD + -dependent aldehyde dehydrogenase (ALDH) to 5-hydroxy-indoleacetic acid (5-HIAA). The blue circles represent 5-HT, black arrows indicate stimulation, and red blunt arrows portray inhibition. (modified from Ning Liu, et al., 2021) [71]. Created with BioRender.com.

activation of the protein kinase enzyme (MAPK) by nitrogen. In primary neurons, there is a potent but transitory activation of the ERK pathway, which operates independently of PKA. The principal route for the activation of neuronal ERKs by 5-HT4 receptors is dependent on Src protein-tyrosine kinase [84].

4. The platelet serotonergic system as a mirror of changes in stress, anxiety, depression, and epilepsy

In relation to the hypothesis of the link between the two locations of the platelet serotonergic system (the brain and peripheral tissues), recent preclinical studies have shown a significant reduction in brain and peripheral 5-HT of rodents (in platelet-rich plasma, or PRP) subsequent to forced swimming. A subchronic treatment with fluoxetine elicited a significant increase in the level of 5-HT in PRP. On the other hand, the same treatment prior to forced swimming led to a lesser decrease in the level of 5-HT in PRP than that observed in the control group of rodents submitted to forced swimming without treatment [85]. Accordingly, the level of peripheral 5-HT is sensitive to forced swimming, fluoxetine treatment, and the combination of the two.

Platelets have been proposed as markers possibly capable of distinguishing between acute and chronic stress as well as between somatic symptoms and psychiatric disorders [86, 87]. The low level of 5-HT in the brain of patients in a depressive state has been regulated by the prescription of SRIs [41, 42]. However, no significant difference was detected in the platelet content of 5-HT or in 5-HT reuptake in patients with post-traumatic stress disorder compared to healthy individuals [88]. Hence, further investigation is needed on the effects of distinct kinds of stress on the dynamics of the platelet serotonergic system.

Interestingly, evidence exists of a greater quantity of plasma 5-HT and a higher density of 5-HT2A receptors on platelets in patients with a combination of major depressive disorder and cardiovascular disease [89, 90]. As can be appreciated, a coherent explanation of the participation of the peripheral serotonergic system in depression has not yet been established. A previous systematic review revealed three different outcomes for the peripheral levels of 5-HT after treatment with antidepressant drugs: an increase, a decrease, and no significant change [91]. As a consequence, the peripheral levels of 5-HT cannot be utilized as a biomarker for the diagnosis of depression or as a measurement of the efficacy of treatments.

Likewise, contradictory results have been found in research on the role of the peripheral serotonergic system among anxiety disorders. For instance, a higher level and activity of SERT have been described in the platelets of patients with anxiety disorders [92]. However, a study on patients with panic disorder discovered a diminished average platelet volume in comparison with the control group of healthy individuals [93].

Despite intense efforts in the last few years to explore the involvement of the platelet serotonergic system in the pathophysiology of neurodegenerative diseases (e.g., epilepsy) [94, 95], the findings have not yet provided a clear picture of the overall contribution of the corresponding elements. On the other hand, there are various reports on adverse effects in peripheral tissues produced by treatments aimed at the platelet serotonergic system in the brain. For example, cabergoline and valproic acid (agonists of 5-HT2A and 5-HT2B receptors used to treat Parkinson's disease and epilepsy, respectively) have been linked to the development of valvular heart disease [96, 97].

5. The platelet serotonergic system in the pathophysiology of coagulopathies associated with viral infections

COVID-19 causes an broad spectrum of diseases due to an exaggerated proinflammatory response, which often progresses to hyperinflammation accompanied by coagulopathy and a procoagulant endothelial phenotype [98]. The close relation between the condition of hypercoagulability and infection by COVID-19 can explain various phenomena observed in clinical practice. For instance, thromboembolic events have been described, such as venous and arterial thrombosis, leading to a greater incidence of strokes, myocardial and cerebral infarction, acute coronary syndrome, and pulmonary embolism [99]. Moreover, the activation of platelet aggregation by endothelial damage and the ensuing interaction with other cells aggravates the potential of hyperinflammation, structural remodeling of the pulmonary vasculature, and cardiovascular disease [100, 101].

5-HT has been shown to decrease the harmful effects of COVID-19 infection [102]. It manifests its activity in macrophages and dendritic cells, suppressing the generation of proinflammatory cytokines and chemokines. Furthermore, 5-HT promotes enhanced cytotoxicity in natural killer cells and reduces the production of tumor necrosis factor alpha (TNF-α) [103], while stimulating the proliferation of endothelial cells by binding to their 5-HT2 receptors [104].

There is evidence that ketanserin (a 5-HT2 antagonist) diminishes pulmonary platelet trapping and inhibits platelet activation and aggregation, thus facilitating respiration and lowering the risk of pulmonary fibrosis and adverse consequences in the kidneys, the CNS, and the cardiovascular system. Additionally, this treatment is able to revert the pulmonary vasoconstriction mediated by 5-HT [105].

SRI antidepressants have been employed to treat COVID-19 patients as well. They prevent the release of inflammatory cytokines [106], present anticoagulant activity [34, 41], and make the immune response more effective through the inhibition of proinflammatory molecules such as TNF-α, IL- ϒ, IL-1β, and IL-6 [107, 108]. A selective SRI, fluvoxamine, is among the most studied and used treatments in the outpatient care of COVID-19 patients with risk factors. It has been shown to be beneficial in treating hypercoagulability in patients with COVID-19, possibly due to the effect that this SRI has on the reduction in platelet 5-HT concentration [109].

Regarding infection with dengue virus, the specific mechanisms involved in the development of thrombocytopenia are unknown. It is believed that platelets and 5-HT play a crucial role in understanding the vascular pathophysiology of this disease [30]. In studies carried out on adult patients with severe dengue, the acute phase induced platelet activation, increased intraplatelet 5-TH concentration, and greater expression of the activated fibrinogen receptor, the lysosomal marker CD63, and P-selectin [110].

6. The platelet serotonergic system in the pathophysiology of acute myocardial infarction

Acute myocardial infarction is the main cause of mortality around the world [111]. The prognosis of patients has been improved by opportune treatment, including pharmacological thrombolysis and percutaneous coronary intervention. However, the restauration of coronary blood flow provokes tissue damage, denominated myocardial reperfusion injury [32].

Platelets play a critical role in response to myocardial injury. Firstly, platelet glycoproteins IIb, IIIa, and VI have an essential function in the adhesion and aggregation of platelets. Secondly, activated platelets recruit proinflammatory leukocytes to the site of ischemic tissue, providing reperfusion during the first few days following a lesion. Thirdly, platelets release exosomes, which increase inflammation within the ischemic myocardium. They also release microvesicles and apoptotic bodies involved in myocardial tissue regeneration. Finally, 5-HT has been found to aggravate inflammation in the area of an infarct [111], suggesting that the respective receptors could be targeted to afford protective effects in post-infarct patients [112]. Some recently developed drugs have been administered to patients with acute myocardial infarction in the early phase of recovery. For instance, ondansetron shows anti-inflammatory effects by modulating the immune system and diminishing platelet activation [112].

7. The platelet serotonergic system in the pathophysiology of preeclampsia

Preeclampsia is a severe hypertensive complication of pregnancy that constitutes a significant risk for the mother and fetus [113, 114]. In women with preeclampsia, the following conditions have been detected: (1) enhanced availability of tryptophan, (2) a decrease in the activity of the kynurenine pathway and thus in the synthesis of indoleamine 2,3 dioxygenase, and (3) a reduction in the degradation of 5-HT by monoamine oxidase type A, resulting in a higher concentration of free 5-HT in maternal circulation [115, 116]. During pregnancy, 5-HT is synthesized in the syncytiotrophoblasts of chorionic villi. With preeclampsia, the generation of 5-HT is significantly lower in the syncytiotrophoblasts, but extracellular 5-HT probably causes placental hyperserotonemia (as found in normal maternal circulation) [117]. Various studies carried out on women with preeclampsia and gestational hypertension demonstrate an increase in 5-HT in the placenta, correlating positively with blood pressure and the severity of the disease [116–118].

Moreover, a release of 5-HT from mastocytes, basophils, and platelets is triggered by lesion or inflammation and can initiate cascades and other immune processes, among which are chemotaxis mediated by 5-HT1A receptors and cellular phagocytosis [119, 120]. Hence, 5-HT is necessary for the normal production of cytokines involved in the activation of inflammatory processes, including IFN-γ, IL-1B, IL-8, IL-12, TNFα, IL-17, and IL-6 [121, 122].

According to multiple reports on women with preeclampsia, the existence of hyperserotonemia during preeclampsia exacerbates several pathological processes, such as endothelial cell damage, platelet aggregation, the development of turbulence in blood flow, and microvascular damage in the placenta, all leading to more abundant inflammation [24, 123]. Damage to endothelial cells promotes platelet aggregation and as a consequence the release of 5-HT, which induces vasoconstriction mediated by 5-HT2 receptors in smooth muscle tissue and the uterine artery [25, 124].

8. Conclusions

The platelet serotonergic system is complex, affecting factors in the CNS and peripheral tissues. SRIs and other serotonergic drugs (agonists and antagonists of 5-HTRs) not only interact with the immune and central nervous system to treat certain diseases, but also may participate in coagulation, vascular permeability, and peripheral vascular resistance. It is crucial to consider both the central and peripheral locations and functions of the platelet serotonergic system before prescribing a medication that targets SERT or a serotonergic receptor (5-HT1A, 5-HT2A, 5-HT3, or 5-HT4) in order to avoid or reduce adverse effects (**Figure 2**). These targets represent an opportunity to treat the pathophysiology of infectious diseases (e.g., dengue and COVID-19), neuropsychological disorders (e.g., stress, depression, anxiety, and epilepsy), and cardiovascular diseases. They also represent potential biomarkers of the same disorders. Further research is required to provide greater insights into the mechanisms at play in the platelet serotonergic system. The challenge in drug development is to discover selective ligands capable of giving rise to the desired activity in the CNS or peripheral tissues (without provoking adverse effects) by increasing or decreasing the activation of platelets.

Conflict of interest

The authors declare that they have no conflict of interest.

Author details

Alfredo Briones-Aranda[1*], Josselin Corzo-Gómez[2], Diana Casique-Aguirre[3,4] and Mauricio Megchún-Hernández[1,5]

1 Facultad de Medicina Humana, Universidad Autónoma de Chiapas, Mexico

2 Escuela de Ciencias Químicas, Universidad Autónoma de Chiapas, Ocozocoautla de Espinosa, Mexico

3 Investigadoras e Investigadores por México CONAHCYT, Mexico

4 Centro de Investigación Biomédica de Oriente, Instituto Mexicano del Seguro Social, Puebla, Mexico

5 Hospital de Especialidades Pediátricas, Tuxtla Gutiérrez, Chiapas, México

*Address all correspondence to: alfredo.briones@unach.mx; alfred725@hotmail.com

References

[1] Thon JN, Italiano JE. Platelets: Production, morphology and ultrastructure. Handbook of Experimental Pharmacology. 2012;**210**:3-22

[2] Ren DW, Liu WJ. Regulation of the megakaryocyte development and the platelet production—Review. Zhongguo Shi Yan Xue Ye Xue Za Zhi. 2018;**26**(6):1876-1880

[3] Hayashi T, Tanaka S, Hori Y, Hirayama F, Sato EF, Inoue M. Role of mitochondria in the maintenance of platelet function during *in vitro* storage. Transfusion Medicine. 2011;**21**(3):166-174

[4] Mohammad-Zadeh LF, Moses L, Gwaltney-Brant SM. Serotonin: A review. Journal of Veterinary Pharmacology and Therapeutics. 2008;**31**(3):187-199

[5] Yubero-Lahoz S, Robledo P, Farré M, de laTorre R. Platelet SERT as a peripheral biomarker of serotonergic neurotransmission in the central nervous system. Current Medicinal Chemistry. 2013;**20**(11):1382-1396

[6] Jolas T, Aghajanian GK. Neurotensin and the serotonergic system. Progress in Neurobiology. 1997;**52**(6):455-468

[7] Erspamer V, Asero B. Identification of enteramine, the specific hormone of the enterochromaffin cell system, as 5-hydroxytryptamine. Nature. 1952;**169**(4306):800-801

[8] Lam DD, Heisler LK. Serotonin and energy balance: Molecular mechanisms and implications for type 2 diabetes. Expert Reviews in Molecular Medicine. 2007;**9**(5):1-24

[9] Wiersma DA, Roth RA. Clearance of 5-hydroxytryptamine by rat lung and liver: The importance of relative perfusion and intrinsic clearance. The Journal of Pharmacology and Experimental Therapeutics. 1980;**212**(1):97-102

[10] Jones LA, Sun EW, Martin AM, Keating DJ. The ever-changing roles of serotonin. The International Journal of Biochemistry & Cell Biology. 2020;**125**:105776

[11] Rendu F, Brohard-Bohn B. The platelet release reaction: Granules' constituents, secretion and functions. Platelets. 2001;**12**(5):261-273

[12] del Río J, Lasheras B. Serotoninergic receptors: Physiological and therapeutic implications. Revista de Medicina de la Universidad de Navarra. 1993;**38**(3):29-38

[13] Bayer H, Müller T, Myrtek D, Sorichter S, Ziegenhagen M, Norgauer J, et al. Serotoninergic receptors on human airway epithelial cells. American Journal of Respiratory Cell and Molecular Biology. 2007;**36**(1):85-93

[14] Zhang ZJ, Wang D, Man SC, Ng R, McAlonan GM, Wong HK, et al. Platelet 5-HT(1A) receptor correlates with major depressive disorder in drug-free patients. Progress in Neuro-Psychopharmacology & Biological Psychiatry. 2014;**53**:74-79

[15] Liu YS, Yang M. The effect of 5-hydroxtryptamine on the regulation of megakaryocytopoiesis. Hematology. 2006;**11**(1):53-56

[16] Spigset O, Wilhelmsson C, Mjörndal T, Eriksson S. Serotonin 5-HT2A receptor binding in platelets

from patients with Alzheimer's disease or vascular dementia. International Psychogeriatrics. 2000;**12**(4):537-545

[17] Stratz C, Trenk D, Bhatia HS, Valina C, Neumann FJ, Fiebich BL. Identification of 5-HT3 receptors on human platelets: Increased surface immunoreactivity after activation with adenosine diphosphate (ADP) and thrombin receptor-activating peptide (TRAP). Thrombosis and Haemostasis. 2008;**99**(4):784-786

[18] Serebruany VL, El Mouelhi M, Pfannkuche HJ, Rose K, Marro M, Angiolillo DJ. Investigations on 5-HT_4 receptor expression and effects of tegaserod on human platelet aggregation in vitro. American Journal of Therapeutics. 2010;**17**(6):543-552

[19] Dib PRB, Quirino-Teixeira AC, Merij LB, Pinheiro MBM, Rozini SV, Andrade FB, et al. Innate immune receptors in platelets and platelet-leukocyte interactions. Journal of Leukocyte Biology. 2020;**108**(4):1157-1182

[20] Harifi G, Sibilia J. Pathogenic role of platelets in rheumatoid arthritis and systemic autoimmune diseases. Perspectives and therapeutic aspects. Saudi Medical Journal. 2016;**37**(4):354-360

[21] Habets KLL, Huizinga TWJ, Toes REM. Platelets and autoimmunity. European Journal of Clinical Investigation. 2013;**43**(7):746-757

[22] Hamzeh-Cognasse H, Berthelot P, Tardy B, Pozzetto B, Bourlet T, Laradi S, et al. Platelet toll-like receptors are crucial sensors of infectious danger moieties. Platelets. 2018;**29**(6):533-540

[23] Stasi R, Willis F, Shannon MS, Gordon-Smith EC. Infectious causes of chronic immune thrombocytopenia. Hematology/Oncology Clinics of North America. 2009;**23**(6):1275-1297

[24] Gumusoglu S, Scroggins S, Vignato J, Santillan D, Santillan M. The serotonin-immune Axis in preeclampsia. Current Hypertension Reports. 2021;**23**(7):37

[25] Stott DJ, Saniabadi AR, Hosie J, Lowe GD, Ball SG. The effects of the 5 HT2 antagonist ritanserin on blood pressure and serotonin-induced platelet aggregation in patients with untreated essential hypertension. European Journal of Clinical Pharmacology. 1988;**35**(2):123-129

[26] Yamamoto M, Jin JJ, Wu Z, Abe M, Tabara Y, Nagai T, et al. Interaction between serotonin 2A receptor and endothelin-1 variants in association with hypertension in Japanese. Hypertension Research. 2006;**29**(4):227-232

[27] Mammadova-Bach E, Mauler M, Braun A, Duerschmied D. Autocrine and paracrine regulatory functions of platelet serotonin. Platelets. 2018;**29**(6):541-548

[28] Ponomarev ED. Fresh evidence for platelets as neuronal and innate immune cells: Their role in the activation, differentiation, and deactivation of Th1, Th17, and Tregs during tissue inflammation. Frontiers in Immunology. 2018;**9**:406

[29] Canobbio I. Blood platelets: Circulating mirrors of neurons? Research and Practice in Thrombosis and Haemostasis. 2019;**3**(4):564-565

[30] Corzo-Gómez J, Picazo O, Castellanos-Pérez M, Briones-Aranda A. Systematic review of the serotonergic system in the pathophysiology of severe dengue: The theory of thrombocytopenia and vascular extravasation. Mini

Reviews in Medicinal Chemistry. 2023;**23**(2):230-243

[31] Santos AP, Couto CF, Pereira SS, Monteiro MP. Is serotonin the missing link between COVID-19 course of severity in patients with diabetes and obesity? Neuroendocrinology. 2022;**112**(11):1039-1045

[32] Ziegler M, Wang X, Peter K. Platelets in cardiac ischaemia/reperfusion injury: A promising therapeutic target. Cardiovascular Research. 2019;**115**(7):1178-1188

[33] Kopeikina E, Ponomarev ED. The role of platelets in the stimulation of neuronal synaptic plasticity, electric activity, and oxidative phosphorylation: Possibilities for new therapy of neurodegenerative diseases. Frontiers in Cellular Neuroscience. 2021;**15**:680126

[34] Halperin D, Reber G. Influence of antidepressants on hemostasis. Dialogues in Clinical Neuroscience. 2007;**9**(1):47-59

[35] Zhu MM, Li HL, Shi LH, Chen XP, Luo J, Zhang ZL. The pharmacogenomics of valproic acid. Journal of Human Genetics. 2017;**62**(12):1009-1014

[36] Johnston JP, Nerenberg SF. Valproic acid-induced thrombocytopenia-related spontaneous systemic bleeding. American Journal of Case Reports. 2020;**21**:e927830

[37] Ni W, Watts SW. 5-hydroxytryptamine in the cardiovascular system: Focus on the serotonin transporter (SERT). Clinical and Experimental Pharmacology & Physiology. 2006;**33**(7):575-583

[38] Linder AE, Ni W, Szasz T, Burnett R, Diaz J, Geddes TJ, et al. A serotonergic system in veins: Serotonin transporter-independent uptake. Journal of Pharmacology and Experimental Therapeutics. 2008;**325**(3):714-722

[39] Brindley RL, Bauer MB, Blakely RD, Currie KPM. An interplay between the serotonin transporter (SERT) and 5-HT receptors controls stimulus-secretion coupling in sympathoadrenal chromaffin cells. Neuropharmacology. 2016;**110**:438-448

[40] Mercado CP, Kilic F. Molecular mechanisms of SERT in platelets: Regulation of plasma serotonin levels. Molecular Interventions. 2010;**10**(4):231-241

[41] Bismuth-Evenzal Y, Gonopolsky Y, Gurwitz D, Iancu I, Weizman A, Rehavi M. Decreased serotonin content and reduced agonist-induced aggregation in platelets of patients chronically medicated with SSRI drugs. Journal of Affective Disorders. 2012;**136**(1-2):99-103

[42] de Abajo FJ, Montero D, Garcia Rodriguez LA, Madurga M. Antidepressants and risk of upper gastrointestinal bleeding. Basic & Clinical Pharmacology & Toxicology. 2006;**98**(3):304-310

[43] Holck A, Wolkowitz OM, Mellon SH, Reus VI, Nelson JC, Westrin Å, et al. Plasma serotonin levels are associated with antidepressant response to SSRIs. Journal of Affective Disorders. 2019;**250**:65-70

[44] Paczkowski FA, Bryan-Lluka LJ, Pörzgen P, Brüss M, Bönisch H. Comparison of the pharmacological properties of cloned rat, human, and bovine norepinephrine transporters. The Journal of Pharmacology and Experimental Therapeutics. 1999;**290**(2):761-767

[45] Berger M, Gray JA, Roth BL. The expanded biology of serotonin. Annual Review of Medicine. 2009;**60**:355-356

[46] De Meyer SF, Vanhoorelbeke K, Broos K, Salles II, Deckmyn H. Antiplatelet drugs. British Journal of Haematology. 2008;**142**(4):515-528

[47] Machida T, Iizuka K, Hirafuji M. 5-Hydroxytryptamine and its receptors in systemic vascular walls. Biological & Pharmaceutical Bulletin. 2013;**36**(9):1416-1419

[48] Gocht A, Distler JHW, Spriewald B, Ramsperger-Gleixner M, Weyand M, Ensminger SM, et al. Effects of different serotonin receptor subtype antagonists on the development of cardiac allograft vasculopathy in murine aortic allografts. Transplant Immunology. 2018;**49**:43-53

[49] Hoyer D, Hannon JP, Martin GR. Molecular, pharmacological and functional diversity of 5-HT receptors. Pharmacology, Biochemistry, and Behavior. 2002;**71**(4):533-554

[50] Ellis ES, Byrne C, Murphy OE, Tilford NS, Baxter GS. Mediation by 5-hydroxytryptamine 2B receptors of endothelium-dependent relaxation in rat jugular vein. British Journal of Pharmacology. 1995;**114**(2):400-404

[51] Borman RA, Tilford NS, Harmer DW, Day N, Ellis ES, Sheldrick RLG, et al. 5-HT(2B) receptors play a key role in mediating the excitatory effects of 5-HT in human colon *in vitro*. British Journal of Pharmacology. 2002;**135**(5):1144-1151

[52] Ramage A, Villalon C. 5-Hydroxytryptamine and cardiovascular regulation. Trends in Pharmacological Sciences. 2008;**29**(9):472-481

[53] Ramage AG. Central cardiovascular regulation and 5-hydroxytryptamine receptors. Brain Research Bulletin. 2001;**56**(5):425-439

[54] Bootle DJ, Adcock JJ, Ramage AG. The role of central 5-HT receptors in the bronchoconstriction evoked by inhaled capsaicin in anaesthetised Guinea-pigs. Neuropharmacology. 1998;**37**(2):243-250

[55] Ramage AG. The role of central 5-hydroxytryptamine (5-HT, serotonin) receptors in the control of micturition. British Journal of Pharmacology. 2006;**147**(S2):S120-S131

[56] Gradin K, Pettersson A, Hedner T, Persson B. Acute administration of 8-hydroxy-2-(di-n-propylamino) tetralin (8-OH-DPAT), a selective 5-HT-receptor agonist, causes a biphasic blood pressure response and a bradycardia in the normotensive Sprague-Dawley rat and in the spontaneously hypertensive rat. Journal of Neural Transmission. 1985;**62**(3-4):305-319

[57] Walstab J, Rappold G, Niesler B. 5-HT3 receptors: Role in disease and target of drugs. Pharmacology & Therapeutics. 2010;**128**(1):146-169

[58] Thompson AJ, Lummis SCR. The 5-HT3 receptor as a therapeutic target. Expert Opinion on Therapeutic Targets. 2007;**11**(4):527-540

[59] Reeves DC, Goren EN, Akabas MH, Lummis SCR. Structural and electrostatic properties of the 5-HT 3 receptor pore revealed by substituted cysteine accessibility mutagenesis*. Journal of Biological Chemistry [Internet]. 2001;**276**(45):42035-42042. Available from: http://www.jbc.org

[60] Derkach V, Surprenant A, North RA. 5-HT3 receptors are membrane ion channels. Nature. 1989;**339**(6227):706-709

[61] Maricq AV, Peterson AS, Brake AJ, Myers RM, Julius D. Primary structure

and functional expression of the 5HT3 receptor, a serotonin-gated ion channel. Science. 1979;**254**(5030):432-437

[62] Parker RMC, Barnes JM, Ge J, Barber PC, Barnes NM. Autoradiographic distribution of [3 H]-(S)-zacopride-labelled 5-HT 3 receptors in human brain. Journal of the Neurological Sciences. 1996;**144**(1-2):119-127

[63] Reeves DC, Lummis SCR. The molecular basis of the structure and function of the 5-HT 3 receptor: A model ligand-gated ion channel (review). Molecular Membrane Biology. 2002;**19**(1):11-26

[64] Fatima SK. Effect of D-serine on the serotonin receptors of human platelets. Experimental Brain Research. 2006;**173**(2):353-356

[65] Nichols DE, Nichols CD. Serotonin receptors. Chemical Reviews. 2008;**108**(5):1614-1641

[66] Kemp A, Manahan-Vaughan D. The 5-hydroxytryptamine 4 receptor exhibits frequency-dependent properties in synaptic plasticity and behavioural metaplasticity in the hippocampal CA1 region *in vivo*. Cerebral Cortex. 2005;**15**(7):1037-1043

[67] Kadowaki M, Wang XO, Shimatani H, Yoneda S, Takaki M. 5-HT4 receptor enhances the propulsive power of the peristaltic reflex in the rat distal colon. Autonomic Neuroscience. 2002;**99**(1):62-65

[68] Gershon MD. Review article: Serotonin receptors and transporters—Roles in normal and abnormal gastrointestinal motility. Alimentary Pharmacology & Therapeutics. 2004;**20**(Suppl. 7):3-14

[69] Irving HR, Tan YY, Tochon-Danguy N, Liu H, Chetty N, Desmond PV, et al. Comparison of 5-HT4 and 5-HT7 receptor expression and function in the circular muscle of the human colon. Life Sciences. 2007;**80**(13):1198-1205

[70] Bockaert J, Claeysen S, Compan V, Dumuis A. 5-HT4 receptors. Current Drug Target-CNS & Neurological Disorders. 2004;**3**(1):39-51

[71] Liu N, Sun S, Wang P, Sun Y, Hu Q, Wang X. The mechanism of secretion and metabolism of gut-derived 5-hydroxytryptamine. International Journal of Molecular Sciences. 2021;**22**(15):7931

[72] Bertrand PP, Bertrand RL. Serotonin release and uptake in the gastrointestinal tract. Autonomic Neuroscience [Internet]. 2010;**153**(1-2):47-57. Available from: https://pubmed.ncbi.nlm.nih.gov/19729349/

[73] Terry N, Margolis KG. Serotonergic mechanisms regulating the GI tract: Experimental evidence and therapeutic relevance. Handbook of Experimental Pharmacology [Internet]. 2017;**239**:319-342. Available from: https://pubmed.ncbi.nlm.nih.gov/28035530/

[74] Dizeyi N, Hedlund P, Bjartell A, Tinzl M, Austild-Taskén K, Abrahamsson PA. Serotonin activates MAP kinase and PI3K/Akt signaling pathways in prostate cancer cell lines. Urologic Oncology: Seminars and Original Investigations. 2011;**29**(4):436-445

[75] Boczek T, Mackiewicz J, Sobolczyk M, Wawrzyniak J, Lisek M, Ferenc B, et al. The role of G protein-coupled receptors (GPCRs) and calcium signaling in schizophrenia. Focus on GPCRs activated by neurotransmitters and chemokines. Cell. 2021;**10**(5):1228

[76] Guillet-Deniau I, Burnol AF, Girard J. Identification and localization of a skeletal muscle secrotonin 5-HT2A receptor coupled to the Jak/STAT pathway. The Journal of Biological Chemistry. 1997;**272**(23):14825-14829

[77] Baldys A, Raymond JR. Role of c-Cbl carboxyl terminus in serotonin 5-HT2A receptor recycling and resensitization. Journal of Biological Chemistry. 2011;**286**(28):24656-24665

[78] Marcinkowska M, Kubacka M, Zagorska A, Jaromin A, Fajkis-Zajaczkowska N, Kolaczkowski M. Exploring the antiplatelet activity of serotonin 5-HT2A receptor antagonists bearing 6-fluorobenzo [d] isoxazol-3-ylpropyl motif–as potential therapeutic agents in the prevention of cardiovascular diseases. Biomedicine & Pharmacotherapy. 2022;**145**:112424

[79] Czopek A, Kubacka M, Bucki A, Siwek A, Filipek B, Pawłowski M, et al. Novel serotonin 5-HT2A receptor antagonists derived from 4-phenylcyclohexane-5-spiro- and 5-methyl-5-phenyl-hydantoin, for use as potential antiplatelet agents. Pharmacological Reports [Internet]. 2021;**73**(5):1361-1372. Available from: https://pubmed.ncbi.nlm.nih.gov/34115343/

[80] Green T, Stauffer KA, Lummis SCR. Expression of recombinant homo-oligomeric 5-hydroxytryptamine 3 receptors provides new insights into their maturation and structure. Journal of Biological Chemistry. 1995;**270**(11):6056-6061

[81] Boess FG, Beroukhim R, Martin IL. Ultrastructure of the 5-Hydroxytryptamine 3 receptor. Journal of Neurochemistry. 1995;**64**(3):1401-1405

[82] Barrera NP, Herbert P, Henderson RM, Martin IL, Edwardson JM. Atomic force microscopy reveals the stoichiometry and subunit arrangement of 5-HT_3 receptors. National Academy of Sciences of the United States of America. 2005;**102**(35):12595-12600

[83] Marumo M, Ekawa K, Wakabayashi I. Resveratrol inhibits Ca^{2+} signals and aggregation of platelets. Environmental Health and Preventive Medicine. 2020;**25**(1):70

[84] Barthet G, Framery B, Gaven F, Pellissier L, Reiter E, Claeysen S, et al. 5-hydroxytryptamine 4 receptor activation of the extracellular signal-regulated kinase pathway depends on Src activation but not on G protein or beta-arrestin signaling. Molecular Biology of the Cell [Internet]. 2007;**18**(6):1979-1991. Available from: https://pubmed.ncbi.nlm.nih.gov/17377064/

[85] Bianchi M, Moser C, Lazzarini C, Vecchiato E, Crespi F. Forced swimming test and fluoxetine treatment: *in vivo* evidence that peripheral 5-HT in rat platelet-rich plasma mirrors cerebral extracellular 5-HT levels, whilst 5-HT in isolated platelets mirrors neuronal 5-HT changes. Experimental Brain Research. 2002;**143**(2):191-197

[86] Sandrini L, Ieraci A, Amadio P, Zarà M, Barbieri SS. Impact of acute and chronic stress on thrombosis in healthy individuals and cardiovascular disease patients. International Journal of Molecular Sciences. 2020;**21**(21):7818

[87] Koudouovoh-Tripp P, Hüfner K, Egeter J, Kandler C, Giesinger JM, Sopper S, et al. Stress enhances proinflammatory platelet activity: The impact of acute and chronic mental stress. Journal of Neuroimmune Pharmacology. 2021;**16**(2):500-512

[88] Cicin-Sain L, Mimica N, Hranilovic D, Balija M, Ljubin T, Makarić G, et al. Posttraumatic stress disorder and platelet serotonin measures. Journal of Psychiatric Research. 2000;**34**(2):155-161

[89] Schins A, Honig A, Crijns H, Baur L, Hamulyák K. Increased coronary events in depressed cardiovascular patients: 5-HT2A receptor as missing link? Psychosomatic Medicine. 2003;**65**(5):729-737

[90] Williams MS, Ziegelstein RC, McCann UD, Gould NF, Ashvetiya T, Vaidya D. Platelet serotonin signaling in patients with cardiovascular disease and comorbid depression. Psychosomatic Medicine. 2019;**81**(4):352-362

[91] Wiaderna D, Gralewicz S, Tomas T. Behavioural changes following a four-week inhalation exposure to hemimellitene (1,2,3-trimethylbenzene) in rats. International Journal of Occupational Medicine and Environmental Health. 1998;**11**(4):319-334

[92] Zhuang X, Xu H, Fang Z, Xu C, Xue C, Hong X. Platelet serotonin and serotonin transporter as peripheral surrogates in depression and anxiety patients. European Journal of Pharmacology. 2018;**834**:213-220

[93] Göğçegöz Gül I, Eryılmaz G, Ozten E, Hızlı SG. Decreased mean platelet volume in panic disorder. Neuropsychiatric Disease and Treatment. 2014;**10**:1665-1669

[94] Somogyi GT, de Groat WC. Evidence for inhibitory nicotinic and facilitatory muscarinic receptors in cholinergic nerve terminals of the rat urinary bladder. Journal of the Autonomic Nervous System. 1992;**37**(2):89-97

[95] Cupello A, Audenino D, Scarrone S, Fornaro M, Gatta E, Fornaro P, et al. Epileptic seizures but not pseudoseizures are associated with decreased density of the serotonin transporter in blood platelet membranes. Neurochemical Research. 2008;**33**(11):2263-2268

[96] Pullan PT. Dopamine agonists, Dopplers and doubt: Cabergoline-associated cardiac valvulopathy. Internal Medicine Journal. 2009;**39**(4):213-215

[97] Solela G, Fedlu M. Rare recurrence of Sydenham chorea in an adult: A case report. International Medical Case Reports Journal. 2023;**16**:265-268

[98] Wool GD, Miller JL. The impact of COVID-19 disease on platelets and coagulation. Pathobiology. 2021;**88**(1):15-27

[99] Monteil V, Kwon H, Prado P, Hagelkrüys A, Wimmer RA, Stahl M, et al. Inhibition of SARS-CoV-2 infections in engineered human tissues using clinical-grade soluble human ACE2. Cell. 2020;**181**(4):905-913.e7

[100] Malerba M, Clini E, Malagola M, Avanzi GC. Platelet activation as a novel mechanism of atherothrombotic risk in chronic obstructive pulmonary disease. Expert Review of Hematology. 2013;**6**(4):475-483

[101] Malerba M, Nardin M, Radaeli A, Montuschi P, Carpagnano GE, Clini E. The potential role of endothelial dysfunction and platelet activation in the development of thrombotic risk in COPD patients. Expert Review of Hematology. 2017;**10**(9):821-832

[102] Szabo A, Gogolak P, Koncz G, Foldvari Z, Pazmandi K, Miltner N, et al. Immunomodulatory capacity of the serotonin receptor 5-HT2B in a subset of

human dendritic cells. Scientific Reports. 2018;**8**(1):1765

[103] Lu B, Kwan K, Levine YA, Olofsson PS, Yang H, Li J, et al. α7 nicotinic acetylcholine receptor signaling inhibits inflammasome activation by preventing mitochondrial DNA release. Molecular Medicine. 2014;**20**(1):350-358

[104] Warner-Schmidt JL, Duman RS. VEGF is an essential mediator of the neurogenic and behavioral actions of antidepressants. Proceedings of the National Academy of Sciences of the United States of America. 2007;**104**(11):4647-4652

[105] Kuindersma M, Spronk PE. Ketanserin as potential additive drug to improve V/Q mismatch in COVID-19? Critical Care. 2020;**24**(1):526

[106] Roumestan C, Michel A, Bichon F, Portet K, Detoc M, Henriquet C, et al. Anti-inflammatory properties of desipramine and fluoxetine. Respiratory Research. 2007;**8**(1):35

[107] Zhang J, Zhang N, Lei J, Jing B, Li M, Tian H, et al. Fluoxetine shows neuroprotective effects against LPS-induced neuroinflammation via the notch signaling pathway. International Immunopharmacology. 2022;**113**(Pt A):109417

[108] Zheng W, Sun HL, Cai H, Zhang Q, Ng CH, Xiang YT. Antidepressants for COVID-19: A systematic review. Journal of Affective Disorders. 2022;**307**:108-114

[109] Reis G, dos Santos Moreira-Silva EA, Silva DCM, Thabane L, Milagres AC, Ferreira TS, et al. Effect of early treatment with fluvoxamine on risk of emergency care and hospitalisation among patients with COVID-19: The together randomised, platform clinical trial. The Lancet Global Health. 2022;**10**(1):e42-e51

[110] Michels M, Alisjahbana B, de Groot PG, Indrati AR, Fijnheer R, Puspita M, et al. Platelet function alterations in dengue are associated with plasma leakage. Thrombosis and Haemostasis. 2014;**112**(08):352-362

[111] Mauler M, Herr N, Schoenichen C, Witsch T, Marchini T, Härdtner C, et al. Platelet serotonin aggravates myocardial ischemia/reperfusion injury via neutrophil degranulation. Circulation. 2019;**139**(7):918-931

[112] Boshen Y, Yuankang Z, Taixi L, Kaifan N, Zhixiang W, Liang L, et al. Effects of ondansetron treatment on outcomes of critically ill patients with myocardial infarction partly through its anti-inflammatory activity. International Journal of Medical Sciences. 2023;**20**(6):709-716

[113] Bolte AC, van Geijn HP, Dekker GA. Pathophysiology of preeclampsia and the role of serotonin. European Journal of Obstetrics, Gynecology, and Reproductive Biology. 2001;**95**(1):12-21

[114] Gumusoglu SB, Chilukuri ASS, Santillan DA, Santillan MK, Stevens HE. Neurodevelopmental outcomes of prenatal preeclampsia exposure. Trends in Neurosciences. 2020;**43**(4):253-268

[115] Santillan MK, Pelham CJ, Ketsawatsomkron P, Santillan DA, Davis DR, Devor EJ, et al. Pregnant mice lacking indoleamine 2,3-dioxygenase exhibit preeclampsia phenotypes. Physiological Reports. 2015;**3**(1):e12257

[116] Ren D, Huang W, Yang M. Immunohistochemical locating and quantitative studies of 5-hydroxytryptamine and 5-hydroxytryptamine receptor in

placental villi of patients with pregnancy induced hypertension. Zhonghua Fu Chan Ke Za Zhi. 1997;**32**(8):480-482

[117] Middelkoop CM, Dekker GA, Kraayenbrink AA, Popp-Snijders C. Platelet-poor plasma serotonin in normal and preeclamptic pregnancy. Clinical Chemistry. 1993;**39**(8):1675-1678

[118] Gujrati VR, Shanker K, Vrat S, Chandravati PSS. Novel appearance of placental nuclear monoamine oxidase: Biochemical and histochemical evidence for hyperserotonomic state in preeclampsia-eclampsia. American Journal of Obstetrics and Gynecology. 1996;**175**(6):1543-1550

[119] Kushnir-Sukhov NM, Gilfillan AM, Coleman JW, Brown JM, Bruening S, Toth M, et al. 5-hydroxytryptamine induces mast cell adhesion and migration. The Journal of Immunology. 2006;**177**(9):6422-6432

[120] Ieni JR, Meyerson LR. The 5-HT1A receptor probe [3H]8-OH-DPAT labels the 5-HT transporter in human platelets. Life Sciences. 1988;**42**(3):311-320

[121] Lichtblau N, Schmidt FM, Schumann R, Kirkby KC, Himmerich H. Cytokines as biomarkers in depressive disorder: Current standing and prospects. International Review of Psychiatry. 2013;**25**(5):592-603

[122] Cloëz-Tayarani I, Petit-Bertron AF, Venters HD, Cavaillon JM. Differential effect of serotonin on cytokine production in lipopolysaccharide-stimulated human peripheral blood mononuclear cells: Involvement of 5-hydroxytryptamine 2A receptors. International Immunology. 2003;**15**(2):233-240

[123] Gonzalez C, Cruz MA, Gallardo V, Albornoz J, Bravo I. Serotonin-induced vasoconstriction in human placental chorionic veins: Interaction with prostaglandin F_2 alpha. Gynecologic and Obstetric Investigation. 1993;**35**(2):86-90

[124] Tanaka N, Nakamura E, Ohkura M, Kuwabara M, Yamashita A, Onitsuka T, et al. Both 5-hydroxytryptamine 5-HT2A and 5-HT1B receptors are involved in the vasoconstrictor response to 5-HT in the human isolated internal thoracic artery. Clinical and Experimental Pharmacology & Physiology. 2008;**35**(7):836-840

Section 2

Enigmas

Chapter 4

Unlocking the Mysteries: Serotonin Receptor Networks Explored

Javeria Tanveer, Ammarah Baig, Rukhsana Rubeen, Shahana Rasheed Qureshi, Nosheen Bashir, Kanza Khan and Kaneez Fatima-Shad

Abstract

Serotonin affects immunological regulation, hemostasis, vasoconstriction, gut motility, and is linked to several diseases. During peristalsis, serotonin (5-HT) is released from the gut mucosa and is largely generated by enterochromaffin cells (ECs) rather than gut microbes. Gut bacteria can stimulate the production of 5-HT. Serotonin in the blood that is retained within the platelets contributes to the production of clots and platelet aggregation. It binds to receptors such as 5HT2A, producing platelet aggregation and neuronal excitement. It regulates vasoconstriction via 5HT1D in cranial blood arteries. Atherosclerosis, thrombosis, and hypertension are some cardiovascular conditions liked to serotonin dysregulation. Serotonin imbalances in the gut influence gut motility and absorption, leading to conditions such as irritable bowel syndrome (IBS). 5-HT receptor subsets (5-HT1, 5HT2B, 5-HT3, 5-HT4, and 5-HT7) in gut are promising therapeutic targets. Serotonin in the Central Nervous System (CNS) controls a variety of behavioral and cognitive activities. 5-HTRs, including 5-HT1A and 5-HT2A, can have conflicting effects on pyramidal neuron firing. The chapter comprehends 5HTRs' involvement in the blood, gut, and brain, emphasizing its significance in modulating a variety of biological activities. Further investigation must be conducted to better comprehend the complexity of serotonin signaling to develop innovative treatment techniques that target serotonin receptor networking.

Keywords: serotonin, vasoconstriction, gut brain axis, 5-HTReceptors, brain disorders, platelets

1. Introduction

Serotonin (5-hydroxytryptamine, 5-HT) plays important roles in humans' central nervous system and the other peripheral systems. In the central nervous system, it acts as a neurotransmitter, controlling brain functions such as autonomic neural activity, stress response, body temperature, sleep, mood, and appetite. In its role as a peripheral hormone, serotonin is unique in controlling the functions of several organs. In the gastrointestinal tract it is important for regulating motor and secretory functions. Apart from intestinal motility, energy metabolism is also regulated by both central and peripheral serotonin signaling. It also has fundamental effects on hemostasis,

IntechOpen

vascular tone, heart rate, respiratory drive, cell growth and immunity. Serotonin regulates almost all immune cells in response to inflammation, following the activation of platelets [1]. Serotonin (5-HT) is a neurotransmitter involved with the regulation of numerous behavioral and biological functions in the body, playing a role in both psychological processes in the central nervous system (CNS) as well as peripheral tissues such as the bone and gut [2]. Serotonin is a monoamine neurotransmitter that plays a role in several complex biological functions [3, 4]. Its common abbreviation is 5-HT because of its chemical name: 5-hydroxytryptamine [5]. The most clinically relevant function of serotonin is in psychiatric disorders; most commonly, its absence appears to be related to depression, anxiety, and mania [6, 7]. The interplay of the 5-HT system with several other classical neurotransmitter systems makes the wide range of brain processes mediated by 5-HT neurotransmission in the CNS more complicated. 5-HT exerts its effects on regulating the neurotransmitter release of these neurons by activating serotonergic receptors on cholinergic, dopaminergic, GABAergic, or glutamatergic neurons [8, 9]. Additionally, 5-HT neurons also engage in co transmission, which is the release of several classical neurotransmitters by a single neuron. Glutamate and perhaps other amino acids are co-transmitters produced by 5-HT neurons, according to research published in [9]. Intense research is being done to better understand the regulation and functional ramifications of this neuronal co-transmission [10].

2. Synthesis, activation and degradation of 5HT in gut

The gastrointestinal (GI) tract, platelets, and the serotoninergic neuronal network of the central nervous system are the primary locations of serotonin (5-hydroxytryptamine, or 5-HT). In addition to being a neurotransmitter, serotonin is a peripheral hormone [1].

Ninety percent of the serotonin in the body is produced by epithelial enteroendocrine (EE) cells, which include enterochromaffin cells (EC) as one of its kinds. Since they serve as sensory transducers and facilitate transepithelial movement, EC cells are distributed throughout the enteric epithelium that runs from the stomach to the colon.

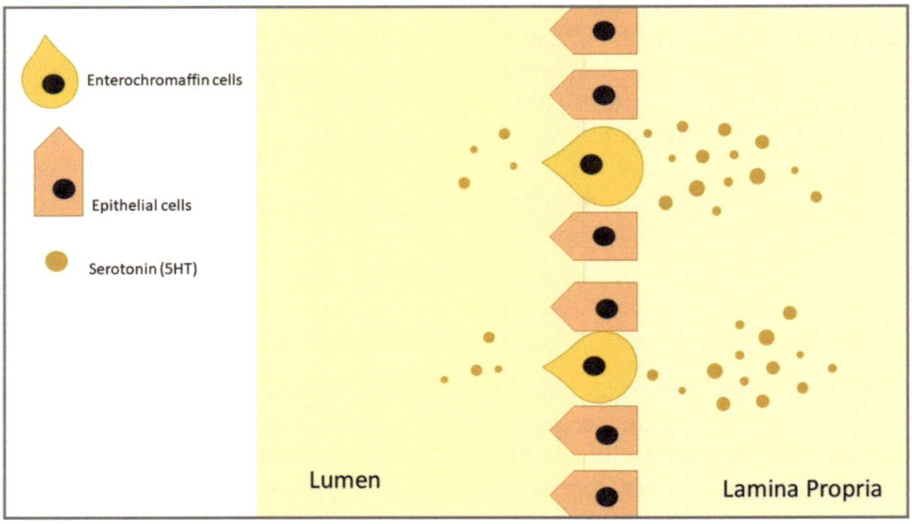

Figure 1.
Serotonin produced in EC cells, with the bulk being released into the lamina propria and a minor quantity into the gut lumen.

This mechanosensitivity is again monitored by 5HTRs. These cells store their 5-HT in secretion granules at the base of the cell, and they have a microvillus border that penetrates into the gut lumen; from where it makes its way to the connective tissue, passing through lamina propria, to gain access to the 5HTRs on the nerve endings, since no nerve ending crosses through the lamina propria [11]. Once synthesized, Vesicular Monoamine Transporter 1 (VMAT1) package and release it from the basal border of EC cell into interstitial space of the mucosa by not only Luminal distention but also by chemical stimuli, which activates of 5HTR in both submucosal and myenteric plexuses on Intrinsic Primary Afferent Neurons (IPANs). The actions of 5-HT are halted by the uptake of serotonin reuptake transporter (SERT), which is expressed by intestinal epithelial cells, platelets, and enteric neurons, into surrounding epithelial cells. It is followed by the intracellular breakdown of 5-hydroxyindoleacetic acid (5-HIAA) by monoamine oxidase (MAO) [12]. Serotonin is released into the intestinal lumen and the lamina propria, which is home to T lymphocytes, dendritic cells, and other immune cells (**Figure 1**) [13].

3. Serotonin receptor networks in brain

Drug reinforcement, stress sensitivity, mood, anxiety, and aggressiveness are all modulated by 5-HT1B receptors. A number of studies have found that decreased 5-HT1B heteroreceptor activation may promote impulsive behavior. The 5-HT1A receptor is a prominent inhibitory G-protein coupled receptor subtype identified in the auto- and heteroreceptor populations of the nervous system. It functions by interacting with Gi/Go proteins, which control a number of intracellular signaling cascades, including cAMP inhibition, calcium channel inactivation, and potassium channel activation [14]. 5-HT5A and 5-HT6 receptors are serotonin proteins present in the human brain that control neurotransmitter release as well as physiological functions such as learning and memory. The HTR5A gene encodes them, and they have been associated with neurological and mental disorders [15].

4. Localization OF 5HT1A, 5HT1B, 5HT6, 5HT5A

4.1 5HT1A

Neurons in the raphe nuclei generate tryptophan hydroxylase 2 (TPH2), which forms the 5-HT system in the brain. Among the 14 5-HT receptor genes, the 5-HT1A receptor is of particular interest because it is abundant in corticolimbic regions implicated in mood and emotion, such as the hippocampal and cortical pyramidal neurons, as well as interneurons of the prefrontal cortex, medial septum, amygdala, hypothalamus, and other regions. On 5-HT neurons, the 5-HT1A receptor is the major somatodendritic auto receptor [16], functioning as a "brake" to lower overall 5-HT system activity and is thought to delay antidepressant response, where it operates as a "brake" to inhibit overall 5-HT system activity and is thought to delay antidepressant response [17]. As a result, processes that control 5-HT1A auto receptor levels are likely to set the tone for the entire 5-HT system [18].

4.2 5HT1B

The basal ganglia, striatum, and frontal cortex are the primary sites of 5-HT1B receptor expression. They are found on presynaptic 5-HT terminals as inhibitory autoreceptors of 5-HT release, it also acts on other nerve terminals as heteroreceptors that govern the release of neurotransmitters such as acetylcholine, glutamate, dopamine, norepinephrine, and gamma-aminobutyric acid [19]. It was recently revealed that the interaction of 5-HT1B receptors with the protein p11 regulates their cellular location. Mice with p11 overexpression and hence have increased 5-HT1B receptor activity have a pattern of antidepressant. p11 mutant mice, on the other hand, display a depression-like phenotype and a reduced responsiveness to antidepressant therapies [20].

4.3 5HT5A

Humans have a protein called HTR5A. It belongs to the 5-hydroxytryptamine receptor protein family. It is a receptor having seven membrane domains that a negative correlation to adenylate cyclase and opens potassium channels when activated [21]. This has been found in the human forebrain, cerebellum, and spinal cord and is encoded by the HTR5A gene [22–24]. It is present in the basal ganglia and the frontal cortex, where it acts as a terminal auto-receptor or heteroreceptor, modulating neurotransmitter release [25].

4.4 5HT6

This is another type found in the human brain [26]. It is a type of protein that regulates how the brain develops. It accomplishes this by interacting with other proteins that regulate how neurons travel and connect to one another. This contributes to the formation of brain circuits critical for memory and behavior [27]. 5-HT6 mRNA is only found in neurons, and it is particularly concentrated near the primary cilia which are 1–5 m long membrane extensions of the cell that are linked to the ciliary basal body. Cilia are important signaling components across the central nervous system. They are found in neurons and are particularly concentrated in the primary cilia. They have also been discovered in numerous locations of the rat brain in connection with both neuronal dendrites and cilia. The subcellular location of 5-HT6 receptors influences their signaling and pathological functions [28].

5. Serotonin receptor dispersion, mechanism, and potential brain impacts

See **Table 1**.

Receptor Type	Location	Role	Route	Action
5-HT1A	hippocampal, Prefrontal cortex pyramidal neurons and interneurons, medial septum, amygdala, and hypothalamus	cognitive function, mood, and emotional states.	Inhibition of ACII	Inhibitory
5-HT1B	basal ganglia, striatum and frontal cortex	mood, memory,	Inhibition of ACII	Inhibitory
5-HT5A	hippocampus and neocortex	learning, memory, and mood regulation.	Inhibition of AC	Inhibitory
5-HT6	cortex and hippocampus	Stress, thinking, retention, and personality	Activation of AC	Stimulatory

Table 1.
Receptor dispersion, mechanism, and effects in the brain.

DOI: http://dx.doi.org/10.5772/intechopen.1004061

6. Pathological implications of serotonin receptors in the brain

The 5-HT1A receptor 5-HT1 inhibitory G-protein-coupled serotonin receptors have critical roles in the clinical manifestations of major depressive disorder, bipolar disorder, schizophrenia, and anxiety-related conditions [29]. In contrast, inhibiting this receptor can cause higher emotional depression in maturity [30] when inhibition of heteroreceptors was commenced in mature individuals, this phenotype was not observed. Inhibition of these receptors causes behavioral despair without any nervousness, whereas aberrant stimulation in this protein during frontal lobe formation restores the whole-brain distressed trait 5-HT1A KO animals [30].

The 5-HT1B receptor When it is expressed in the presynaptic membrane, stimulation of the 5-HT1B. The serotonergic receptors enhance calcium and potassium channel regulation, increasing potassium conductivity, stimulating the cell, and thus

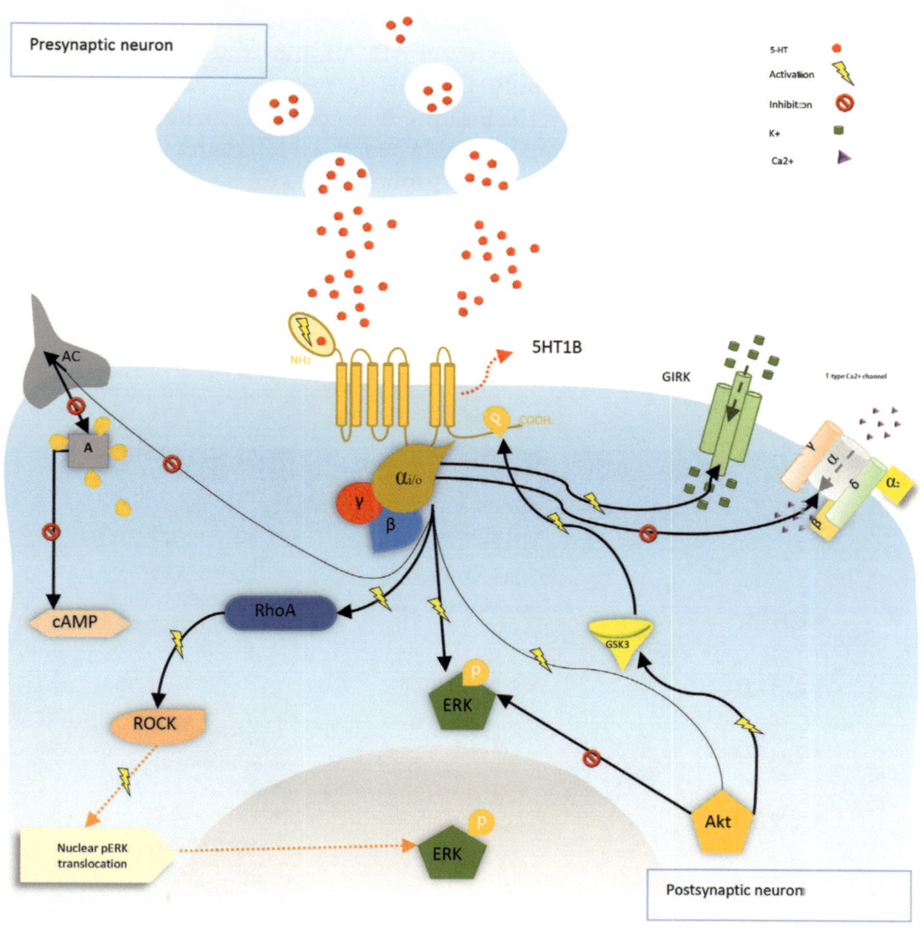

Figure 2.
The activation of 5-HT1B inhibits adenylate cyclase (AC), decreasing ATP conversion to cyclic adenosine Monophosphatase (cAMP) which activates protein kinase a (PKA), the stimulation of these specific receptors reduces Ca2+ and increases K+ conductance. A chain reaction of kinases governs the extracellular signal-regulated kinase (ERK) transfer after 5-HT1B activation. Furthermore, kinase B protein (AKT) is activated, resulting in the stimulation of glycogen synthase kinase 3 (GSK3), which is implicated in the phosphorylation and control of 5-HT1B receptor function.

suppressing 5-HT release. 5-HT1B, as a post-synaptic receptor, suppresses AC via the signaling pathway induced by the Gi/0 protein. When 5-HT1B is activated, it activates ERK, AKT, and the Rho-kinase (ROCK) pathway, which causes activated ERK to be transported to the nucleus. The glycogen synthase kinase 3 protein (GSK3) was linked to putative phosphorylation of the receptor, boosting its repressive capability on AC as a strategy to modulate this signaling pathway (**Figure 2**) [31, 32].

The *5-HT5A* is a widely expressed serotonin receptor in the central nervous system [30, 33]. Despite the receptor's role is not entirely understood, research has suggested that it may be involved in mental illnesses such as schizophrenic and unipolar depressive disorders [23, 30, 34]. In particular, the selective ligand for 5-HT5A is useful in the therapy of cognitive or mood disturbance associated with these disorders However, the information available on specific disorders involving this receptor is currently limited [14].

The *5-HT6 receptor* has been linked to the onset of Alzheimer's illness (AD) and cognitive decline [24]. The role of 5-HT6 in Alzheimer's disease pathogenesis is still not fully understood, but this receptor is thought to be engaged in the regulation of hippocampal cilia function in animal models of AD. Additionally, one study found that the density of 5-HT6 receptor-binding sites in the brain was reduced in patients with AD, and that hyperactivity was the best predictor of how much decreased 5-HT6 in the brain. 5-HT6 receptor inhibitors' possible beneficial effects on therapy in AD suggest that targeting this receptor may hold promise for the generation of novel ideas for treatments in order to improve neurological dysfunction in patients of AD [25].

Alzheimer's disease (AD) is one of the leading reasons for mortality and becoming disabled globally, with severe clinical and socioeconomic consequences. 5-HT6R has been recommended to be a possible pharmacological target for cognitive

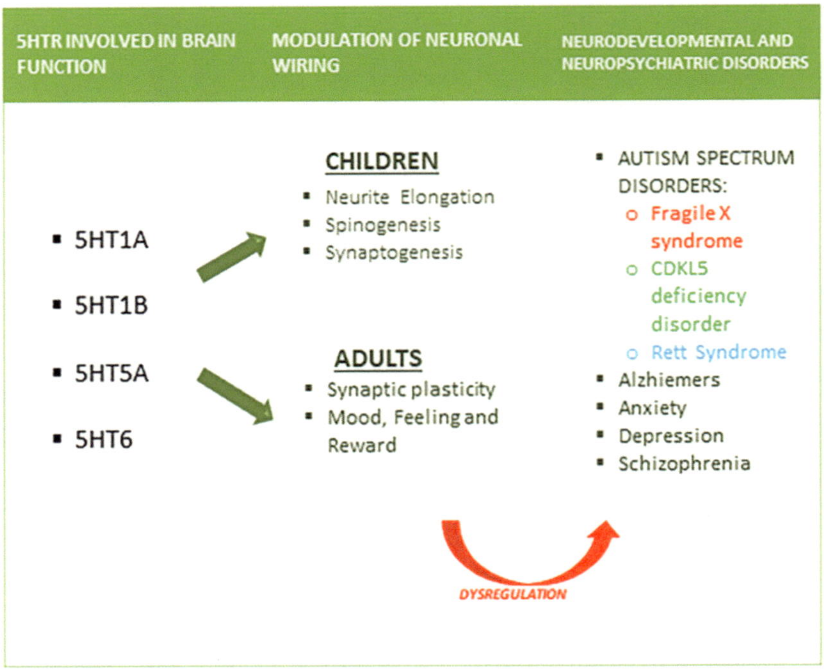

Figure 3.
Implications of dysregulation of 5HTR in brain functions in adults and children.

improvement in Alzheimer's disease in the hunt for innovative treatment options [35]. G protein-coupled receptors, promote adenylate cyclase activity, increasing cAMP production when activated by an agonist. It has been linked to Ca2+ signaling and act on Fyn-tyrosine kinase, a member of the Src family. These findings suggest that 5-HT6 receptors activate ERK1/2 through a Fyn-dependent mechanism. Tau, A microtubule-associated protein that is essential for the formation of tangled neurofibrillary fibers and AD. Mutations in 5HT6 receptors may lead to enhanced tau phosphorylation (**Figure 3**) [26].

7. Therapeutic target of serotonin receptors

The 5-HT6R represents one of the newest receptors for serotonin to be discovered and has attracted a lot of interest following studies that showed it has pro-cognition features. 5HTR5A receptor proteins assist in regulating how the brain functions, such as how it recalls things and how it reacts to particular stimuli. They also aid in the circulatory control of the body. Agonists and antagonists are medicines that can influence how the 5-HT5A receptor works. Serotonin uptake inhibitors are medications that can help the brain recall things better [36]. This protein aids in the transmission across the nervous system and is involved in conditions like depressive illness and pain [37].

It participates in a number of physiological processes, including, memory and learning, and is connected to several neurological and psychological illnesses Drugs that inhibit this receptor may be used to treat memory issues or Alzheimer's disease, 5-HT6R has been a good target for treating memory deterioration, owing to its modulatory role in cholinergic and glutamatergic systems. Blocking the 5-HT6R improves Brain functions in adult rodents, alters age-associated and pharmacologically induced impairments, and can assist in the repair of brain damage induced by early life experiences. This could lead to the creation of unique treatments for mental health issues Neuronal plasticity may benefit from 5-HT6R inhibition as well. Numerous 5-HT6R inhibitors are now being explored in clinical trials in response to these findings [38, 39].

8. Serotinin receptor network in blood

Enterochromaffin cells produce serotonin (5-HT), which is absorbed by platelets in the circulation through the serotonin transporter. Upon secretion by platelets, serotonin triggers both platelet aggregation and vasoconstriction at bleeding sites. Through receptors on the plasma membranes of vascular smooth muscle tissues, blood vessels, and neurons, this signaling process takes place [27].

Smooth muscle cell surface receptors like 5-HT1B and 5-HT2A are the main conduits for the vasoconstrictive actions of serotonin. However, 5-HT1B receptors are the main pathway via intracranial arteries express vasoconstriction. Additionally, the brain hosts the 5-HT1D receptor subtype, concentrated in regions linked to migraine headaches and pain modulation [40]. Activating 5-HT1D receptors can induce vasoconstriction in specific brain areas, offering potential relevance for migraine treatment.

Additionally, serotonin acts as a vasodilator by inducing endothelial cells to release nitric oxide. Different from the actions of 5-HT2A receptors, 5-HT1B and 5-HT1D transmitters primarily control this response [41]. The precise distribution

Serotonin receptors	Location	Main function	Potential effect
5-HT2A	smooth muscle cells of blood vessel, Platelet	Regulating vascular tone and blood flow	Platelet Aggregation, Contraction
5-HT1B	Cranial Blood Vessel	Smooth muscle contraction.	Autoreceptor, Vasoconstriction
5-HT1D	Presynaptic Neuron	Constriction of intracranial blood vessel smooth muscle	Vasoconstriction

Table 2.
Receptors distribution, mechanism of action, and potential effects in blood.

of 5-HT receptors in various vascular smooth muscles, nearby vessel tissue, and the complex regulatory system controlling vascular tone, which includes parasympathetic nerves, determine whether serotonin has a vasoconstrictor or vasodilator effect; see **Table 2**.

The management of serotonin receptors and their activities in the blood is a complicated and closely regulated process. Similar to any other physiological system, abnormalities in serotonin signaling can significantly affect a number of physiological processes and may contribute to the development of particular disorders.

8.1 5-HT2A

5-HT2A receptors (stands for 5- hydroxytryptamine) belongs to the Gg-coupled protein receptor (GCPR) which perform their role in several intracellular pathways. The central nervous system (CNS) has the highest quantity of 5-HT2A mRNA and protein, while blood cells, lymphocytes, and smooth muscle cells in the vascular system all express 5-HT2A. The main roles of 5-HT2A in the vascular system is inflammation and wound repair. For instance, serotonin is released when the endothelium of a blood artery is injured, and smooth muscle cells in the resistance vasculature are activated to produce vasoconstriction [42].

8.2 Molecular mechanisms and functions of 5-HT2A receptors

Vascular smooth muscle cells have a 5-HT2A receptor on their surface, which serotonin binds to. The 5-HT2A receptor's structure experiences a conformational shift that enables it to exchange guanosine triphosphate (GTP) for guanosine diphosphate (GDP), enabling interaction with a particular class of G protein known as Gq. Phospholipase C [43] is activated by the Gq protein, and it cleaves phosphatidylinositol 4,5-bisphosphate (PIP2), a membrane phospholipid, into its secondary carriers: imidazol trisphosphate (IP3) and diacylglycerol (DAG).

When IP3 attaches to its receptor on the membrane of the endoplasmic reticulum, it diffuses into the cytoplasm and releases calcium ion (Ca2+) from intracellular storage. A higher amount of calcium in the cytoplasm activates signaling molecules and calcium-dependent kinases, which phosphorylates and activates myosin light chain kinase (MLCK). Vascular cells of smooth muscle contract as a result of a series of processes that are started when active MLCK phosphorylates the light chains of myosin (MLC). Vasoconstriction and an increase in vascular resistance are the results of this contraction [44].

When blood vessels constrict due to 5-HT2A receptor activation, the reduced blood flow and increased shear stress can lead to molecules that are released from the injured endothelium and the surrounding tissues. One of these molecules is adenosine diphosphate (ADP), which is stored in platelet granules [28, 45]. Activated platelets produce ADP and serotonin, which attach to the appropriate receptors on the platelets, triggering platelet activation. Platelets become sticky and undergo shape change. The surface of carbohydrate IIb/IIIa receptors on platelets become visible upon activation. Blood-circulating fibrinogen molecules have the ability to attach to nearby platelets' activated glycoprotein IIb/IIIa receptors. However binding of fibrinogen bridges adjacent tissues and platelets, causing them to adhere to each other and form aggregates or clumps. This process is known as platelet aggregation. The platelets that have accumulated create a temporary obstruction at the location of the vessel damage. This obstruction helps to prevent excessive bleeding and initiate the process of blood clotting [45].

8.3 5-HT1B

5-HT1B receptors are found in large quantities all over the body. The 5-HT1B receptors are found on serotonergic neurons in the brain and spinal cord, where they control the release of 5-HT from the end of the nerve itself (autoreceptors); [46] because these receptors are primarily translocated to the terminals of axons, there is an anatomical inconsistency between the location of competent 5-HT1B receptor protein and mRNA. It has been discovered that 5-HT1B autoreceptors decrease 5-HT production and release while increasing reuptake through the serotonin transporter. Selective 5-HT1B (/1D) agents, also known as triptans, are utilized as antimigraine medications because 5-HT1B receptors are also present on the smooth muscles membrane of arteries and control 5-HT-induced vasoconstriction across all blood vessels [28]. Additionally, 5-HT1B receptors may facilitate constriction of blood in intra- and extracranial arteries. Although 5-HT1D and 5-HT1B receptors were formerly believed to be the rat equivalents of each other, it is now known that these two receptors are found in every mammalian species investigated and have different geographical distributions. The primary function of the closely related 5-HT1B and 5-HT1D transmitter subtypes is to suppress the production of neurotransmitters like as norepinephrine, dopamine, and serotonin. Depending on the cell types expressing these subtypes, this can have an impact on mood, pain perception, and other neurological functions [47].

8.4 5-HT1D

The central nervous system (CNS), specifically areas of the brain like the cerebral cortex, the hypothalamus and trigeminal nerve pathways, is home to an additional variant of the serotonin receptor, 5-HT1D. Human migraine neurons and trigeminal nerves have been shown to have both 5-HT1B and 5-HT1D receptors; however, only 5-HT1D receptor have been found in trigeminal nerves that bulge externally to the dural endothelium and centrally to the spinal cord trigeminal nuclei [48]. Thus, the 5-HT1D receptors are localized centrally to block the transmission of pain signals from blood vessels to brainstem sensory neurons, as well as superficially to restrict stimulated sensory nerves and prevent the production of vasoactive neuropeptides [41]. It is especially pertinent to processes pertaining to discomfort adjustment and migraine headaches because of its distribution.

9. 5HT1D/1B mediated vasoconstriction

The interactive relation between the receptors of 5-HT1B and 5-HT1D with the context of vasoconstriction involves a presynaptic modulation of the vasoconstrictive response through the inhibition of serotonin release. These receptors, which belong to the family of the G-protein coupled receptor (GPCR) family, carry signals from external ligands to intracellular signaling pathways, including hormones and antidepressants. On serotonergic terminals of the brain, the 5-HT1D receptor is primarily found initially [43]. Upon activation, it exerts inhibitory control over serotonin release into the synaptic cleft, thereby indirectly influencing vasoconstriction mediated by the 5-HT1B receptor subtype. Upon serotonin release from serotonergic nerve terminals, it binds to postsynaptic 5-HT1B receptors on smooth muscle cells a conformational change occurs, which causes a G-protein connected to the ligand (usually a Gq/11 protein) to become activated. Smooth muscles Vasoconstriction can be the activated by Gq/11 proteins that are already defined in section. Simultaneously, a portion of the released serotonin binds to initially 5-HT1D receptors located on the same serotonergic nerve terminals. This binding event activates a signaling cascade, leading to the attenuation of further serotonin release from the nerve terminal. The 5-HT1D receptor experiences a conformational change upon serotonin binding, this results in the stimulation of a class of proteins called Gi/o-type G-proteins, starting transmission cascades downstream [49]. Guanosine diphosphate (GDP) linked to the alpha subunit of Gi/o-type G-proteins is exchanged for guanosine triphosphate (GTP) when serotonin binding activates the 5-HT1D receptor. Gi/o-type G-protein activation results in the dissociation of the alpha subunit (5-HT1Dα) from the beta-gamma fractions (5-HT1Dβγ), which in turn modifies cellular signaling pathways. Gi/o-type G-proteins are triggered by the 5-HT1D receptor to inhibit adenylyl cyclase and reduce the quantity of cyclic AMP (cAMP) generated from ATP (also known as adenosine triphosphate). Consequently, the cAMP-dependent protein kinase (PKA) expression is reduced [50], which results in less phosphorylation of targeted prostate specific antigen. Based on the connective tissue and cell type, the activation of Gi/o-type G-proteins can also open or shut a few ion channels. As a result, the membrane potential and calcium levels fluctuate, changing the release of neurotransmitters and the electrical activity of neurons. The reduced release of serotonin, driven by 5-HT1D receptor activation, results in a diminished availability of serotonin in the synaptic cleft. Therefore, there is a decreased binding of serotonin to postsynaptic receptors, including the receptors of 5-HT1B on the cells of smooth muscle.

The 5-HT1D receptor functions as an indirect regulator of vasoconstriction refers by the receptor of 5-HT1B to modulating serotonin composition ratio which is released from serotonergic nerve terminals **Figure 4(a)**. Through this inhibitory control, the 5-HT1D receptor fine-tunes the overall vasoconstrictive effect of serotonin on smooth muscle cells, which in turn influences vascular tone [23] and blood vessel diameter. This complicated interplay relation within the receptors of the 5-HT1D and 5-HT1B, to plays a crucial role for maintaining physiological balance in the vascular system. 5-HT1B receptors induce vasoconstriction during migraine attacks, potentially reducing blood flow and alleviating pain associated with vascular dilation. On the other hand, 5-HT1D receptors modulate serotonin release resulting in a decreased release of neuropeptides that promote inflammation and contributing to overall vascular constriction during migraines. Triptan medications, which target both receptor subtypes, exploit these vasoconstrictive effects to provide relief for individuals experiencing migraines.

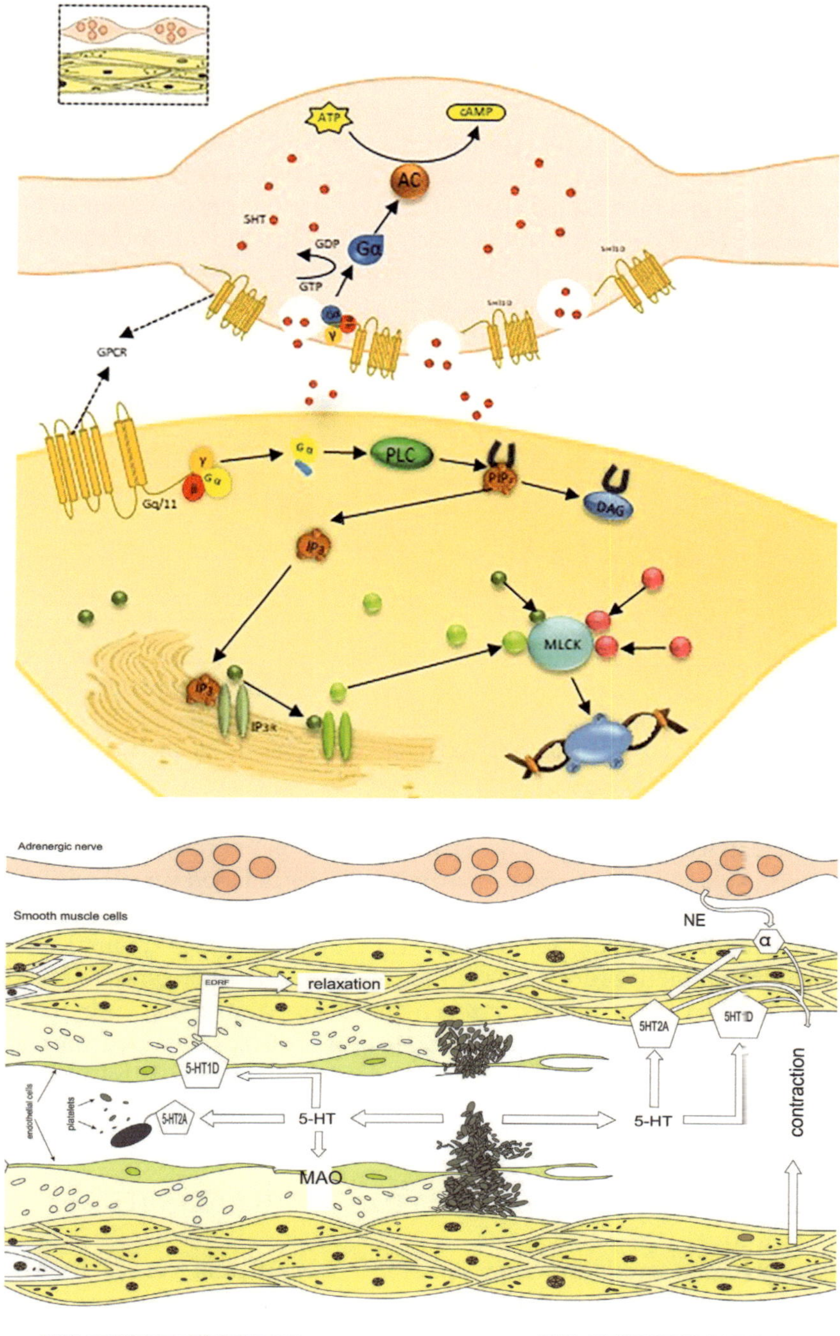

a

b

Figure 4.
(a) Mechanism of 5HT1B/ID mediated vasoconstriction. 5-HT1D receptor functions as an indirect regulator of vasoconstriction mediated by the 5-HT1B receptor by modulating the amount of serotonin released from serotonergic nerve terminals. Through this inhibitory control, the 5-HT1D receptor fine-tunes the overall vasoconstrictive effect of serotonin on smooth muscle cells, which in turn influences vascular tone and blood vessel diameter. (b) Mechanistic action of vasocontraction and dilation depicting the communication between nerve terminals and smooth muscle cells.

10. 5HT1D/1B mediated vasodilation

The free serotonin in the plasma in the systemic circulation is derived from production by chromaffin cells (primarily in the gastrointestinal tract) and overflow into the venous blood from serotonergic neuroeffector junctions (primarily in the brain). The majority of the monoamine is quickly absorbed by platelets (where it is stored in dense granules) and endothelial cells (where it is primarily metabolized by monoamine oxidase (MAO)) thanks to the action of the serotonin transporter (SERT). This ensures a relatively low plasma level of serotonin under physiological conditions [23].

On the other hand, a significant rise in the local concentration of serotonin is unavoidable whenever platelets congregate close to the blood vessel wall and release their dense granules [27]. Additionally, if the monoamine diffuses to the adventitia, it may potentially impact the activity of the sympathetic nerve endings. This activation of serotonergic receptors can occur in both endothelial and vascular smooth muscle cells.

The release of serotonin (5-HT) from aggregating platelets induces more platelet aggregation; at this point unaggregated platelets readily absorb and remove it from the plasma. The monoamine is also taken up by the endothelial cells if the endothelium is intact, and monoamine oxidase (MAO) breaks it down there [51]. Lastly, the endothelium functions as a physical barrier that prevents serotonin-producing vasoconstrictor platelet products from entering the smooth muscle. The endothelium-derived relaxing factors (EDRFs; primarily nitric oxide) **Figure 4(b)** are released when the 5-HT1D and receptors on the endothelium cells are activated. These EDRFs diffuse to the underlying vascular smooth muscle and relax them [23], opening the blood vessel and subsequently flushing the microaggregate away.

In addition, nitric oxide will be released into the lumen to prevent platelets from adhering to the endothelium and to work in concert with prostacyclin to prevent additional platelet aggregation. These various endothelium functions are essential in preventing blood clotting and vasospastic episodes in blood vessels with a healthy intima [41].

When endothelial cells are eliminated (for example, through trauma), the endothelium's protective function is lost locally. Serotonin then binds to the 5-HT2A receptors in the vascular smooth muscle cells, causing contraction via both direct activation of the cells and an increase in their sensitivity to other vasoconstrictors (such as norepinephrine (NE) acting on α1-adrenoceptors (α)). This vasoconstriction then occurs, which plays a part in the vascular phase of hemostasis [43, 52].

11. Pathological implications of 5-HT1D, 5-HT1B AND 5-HT2A

Disturbances in serotonin receptors 5-HT1D, 5-HT1B, and 5-HT2A have been implicated in various pathological conditions affecting the blood. Dysregulation of these receptors can lead to transformed vascular tone, platelet function, and overall hemostasis In the case of 5-HT1D and 5-HT1B receptors, their activation often results in vasoconstriction, which can lead to an increased risk of hypertension and impaired blood flow. This heightened vasoconstrictive response can contribute to conditions such as migraine, where aberrant 5-HT1D and 5-HT1B receptor activity may trigger intense vasoconstriction and subsequent dilation, leading to the characteristic headache [53]. Additionally, disturbances in 5-HT1B receptors have been associated with an increased susceptibility to thrombotic events, as these receptors play a role in platelet aggregation [54]. When overstimulated, 5-HT1B receptors can lead to excessive platelet activation and adhesion, potentially resulting in the formation of pathological blood clots, contributing to conditions such as ischemic stroke or deep vein thrombosis. Similarly, disruptions in 5-HT2A receptor function can impact platelet aggregation and vasoconstriction. Dysfunctional 5-HT2A receptors may play a role in the development of atherosclerosis [52] and other cardiovascular diseases where abnormal platelet function and endothelial dysfunction are key pathological factors. Therefore, understanding the intricate role of these serotonin receptors in blood-related processes is crucial for unraveling the underlying mechanisms of various cardiovascular pathologies and informing potential therapeutic strategies [55].

12. Therapeutic target of 5-HT1D, 5-HT1B AND 5-HT2A

There is still much to learn about the pathogenesis of migraines. Nonetheless, it is thought that dilation of the extracranial, dural, and/or pial arteries plays a key role in the development of migraine headaches. Because of triptans' effectiveness on 5-HT1B receptors, these selective agonists of 5-HT1B/1D receptors were created as cranial vasoconstrictors. The first commercial triptan preparation, sumatriptan, caused vasoconstriction in both human isolated middle meningeal and temporal arteries. However, the selective 5-HT1B receptor antagonist SB224289.40 prevented this effect. On the other hand, sumatriptan-induced vasoconstriction of these arteries was not significantly affected by BRL15572, a selective antagonist of the 5-HT1D receptor [56].

NO-releasing Triptans e.g., Zolmitriptan-NONOate a chemical compound, a newer class of medications designed to combine the vasoconstrictive effects of triptans with the vasodilatory effects of nitric oxide (NO). These compounds act as antagonists at 5-HT1B/1D receptors while simultaneously releasing NO, which counteracts vasoconstriction and promotes vasodilation. This dual mechanism aims to provide more balanced and potentially better-tolerated relief for migraine sufferers. Serotonin (5-HT) itself acts as an endogenous agonist at the 5-HT2A receptor [57]. When released from platelets and other sources, serotonin can bind to 5-HT2A receptors on platelets and potentially contribute to platelet aggregation. While the exact mechanisms are complex and involve multiple receptor subtypes, the activation of 5-HT2A receptors may participate in platelet aggregation and thrombus formation. Ketanserin is a medication that acts as an antagonist at the 5-HT2A receptor. By blocking the binding of serotonin to these receptors, Ketanserin may inhibit or reduce the platelet activation and aggregation response mediated by 5-HT2A receptor [58]. Ketanserin's antagonistic activity at 5-HT2A receptors has been explored for its potential therapeutic effects, including its influence on platelet function and vascular health.

13. Serotonin receptor network in bowel

The main source of 5-HT in the body is the gut and receptors for 5-HT in the gut are the key to regulate various physiological and processes such as colonic motility, secretion, visceral sensitivity and inflammatory/immune response. Such receptors that are localized in the gut are; 5HTR2B, 5HTR3, 5HTR4 and 5HTR7 **Figure 5**. The diverse effects are achieved through their activation, while their localization achieves specific actions [59].

The enterochromaffin cells are the whole source of serotonin production and storage which also express 5HTRs. Together with these 5HTRs, it functions as a

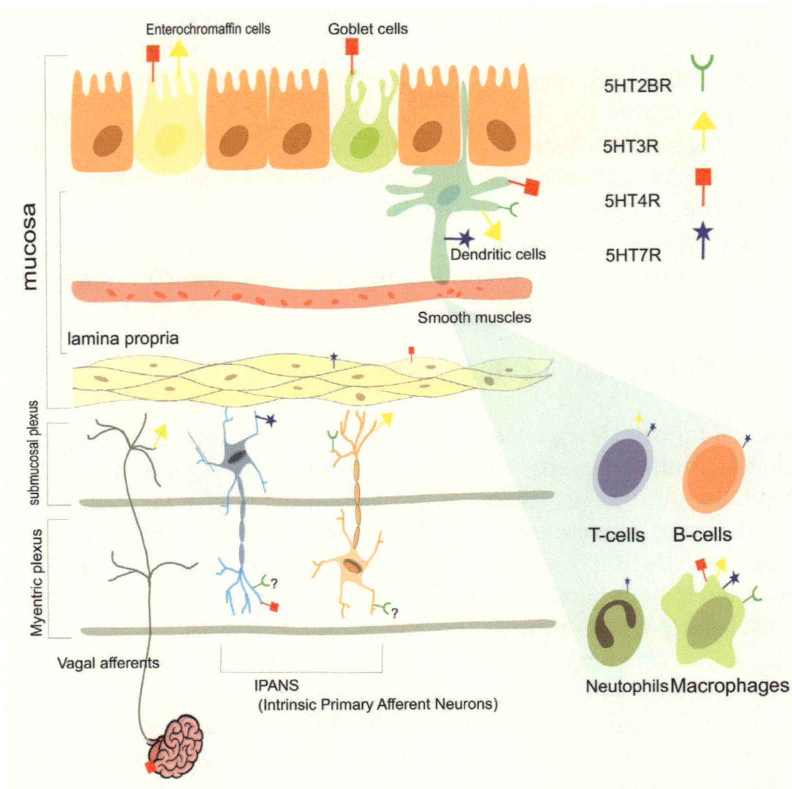

Figure 5.
Serotonin receptors presentation on various layers of the GUT with the cells involved.

neurotransmitter in a bi-directional manner which facilitates the gut-brain axis. This interaction of 5HT with 5HTRs allow continuous information exchange between organs in the GI tract through neural pathways and immune signals where their components also express these receptors [60].

14. 5HTR mediated serotonin functions in GIT

5HT- Serotonin plays a diverse role in modulating various metabolic functions in the body by binding with certain receptors. A vast 5-HT signaling system functions for the gastrointestinal tract (GIT). 5HT, be it synthesized by the EC cells or the gut microbiota, [61] it performs an array of functions as in **Table 3**.

It triggers mucosal peristaltic reflex through interactions with 5-HTR4, 5HT2, 5HT3, 5HT4, and 5HT7 subtypes, impacting gut motor function. It drives propulsive and segmentation motility via 5HT3 and 5HT4. Additionally, 5-HT induces relaxation via 5HT7 and 5HT4 receptors on smooth muscle cells. It enhances epithelial secretion, fat absorption, and vasodilation, while retarding gastric emptying through the receptors 5-HT3. It activates afferent vagal nerve endings, that lead to discomfort and nausea. 5-HT also elicits neurogenic secretory responses through 5-HT2, 5-HT3, and 5-HT4 receptors, supporting content neutralization. It enhances bile acid synthesis and influences glucogenesis affects glycogen synthesis. In the pancreas, it governs

	5HT2B	5HT3	5HT4	5HT7
Locations [62]	Stomach fundus, smooth muscle of the small intestine. Human colon, on longitudinal smooth muscle and the myenteric nerves	Peripheral neuron, CNS	All over the body, Smooth muscles in GIT and neurons of the nervous system in the intestine.	Smooth muscle of vascular and nonvascular tissues including – stomach, ilium, colon, with less representation seen in liver, kidney, spleen and central nervous system.
Agonists	α-Me-serotine BW723C86 Iodoamphetamine-2,5-dimethoxy-4, Quipazine BW-723C86 RO600175 WAY-161503	MKC-733 2-Me-5-HT CPBG Pumosetrag	Tegaserod Prucalopride Cisapride	5-CT 8-OH DPAT
Antagonists	SB200646 SB204741 SB206553 RS 127445 Methysergide Ritanserin	Ondansetron Alosetron Cilansetron	GR113808 GR125487 Piboserod	Methiotepin Metergoline SB258719 SB269970 SB656104
Effectors	Gq/11	Ion channel in Ligand gates	GS	GS
In the gut distribution	Longitudinal smooth muscle	Enteric neurons, EC cells	In the enterocytes, neurons of intestine, and smooth muscles of the enterocytes	In the enterocytes, neurons of intestine, and smooth muscles.
On stimulation-Functional response	Contraction	Increased secretion and transmitter release	Increased transmitter release, and secretion. Smooth muscles relaxation	Smooth muscles relaxation
Immune cells	Macrophages and Dendritic cells (DCs)	T and B cells, DCs, and Macrophages,	Macrophages, DCs,	Neutrophils, T and B cells, DCs, and Macrophages,

Table 3.
Major serotonin/5-hydroxytryptamine/ 5-HT receptor in the bowel -types and subtypes and their location, distribution, functional response, related immune cells and effector agonists and antagonists.

insulin secretion and postprandial pancreatic protein release via 5HT3 and 5HT2 receptors. Furthermore, serotonin molecules also act for pro-inflammatory response, contributing to gastrointestinal pathophysiology, engaging immune cells and by activating serotonin receptors on DCs in the lamina propria [63, 64]. Even if all kind of serotonin receptor have potentiality of activation by the serotonin, as referenced in the **Table 3** above, there is differences in anatomical location, specific distribution in enteric nervous system, harmony for synthetically similar kind of molecules and signal-transduction mechanisms, that make each subtype of serotonin receptor a probable curing and healing target [65].

14.1 5-HT2B

There are three members in the serotonin receptor family of 5-HT2, i.e. 5-HT2A, B and C subtype receptors. 5-HT2B receptor have its importance coupled with the family of G-protein-related receptor. These are heavily expressed in the gastrointestinal tract, as well as in the liver, kidney, and heart. The 5-HT2B receptors in the GI tract the are found in the fundus of stomach, small intestinal smooth muscles, also expressed in intestinal neurons and as well as in the muscles. They are highly expressed in human colon, with predominant localization on longitudinal smooth muscle and the myenteric nerves that control its motility [66]. It Influences smooth muscle rhythmic contraction to propel food forward via peristalsis and secretions production. In interstitial cells of Cajal (ICC), 5-HT2B receptor modulates motility. That expression of 5-HT2B receptors in the ICC network also, operate as a GI "pacemaker" that suggests its role in the smooth muscle contraction. Network of ICC if dysregulated may lead to sluggish intestinal motility [67]. They may also serve as morphogenic agents during the development of enteric neurons by binding with endogenous 5HT in embryonic days, their agonist's increases the enteric neurons differentiation in vitro [62, 67].

14.2 5HT3

The 5-HT3 receptor stands out among all known serotonin receptor subtypes in that it belongs to the ligand-gated ion channel superfamily and is homomeric in nature. This receptor superfamily is comprised of up of five subunits, each featuring four transmembrane segments and a large extracellular N-terminal region. They are expressed on excitable neurons of the gut, notably intrinsic afferent nerves that radiate into the mucosa, interneurons, inhibitory and excitatory motor neurons, ICCs, smooth-muscle cells, and the enterocytes [68]. 5-HT3 stimulation triggers both intrinsic and extrinsic afferent neurons as well as a modest number of excitatory postsynaptic potentials (EPSPs). Neurons in the myenteric plexus, which function as primary afferent neurons and project to the mucosa, undergo stimulation directly by 5-HT presented to the mucosa, where the response is exclusively mediated by 5HT3. They have the capacity to regulate motility, intestinal secretion, visceral sensitivity, and the emetic pathway. 5-HT3 receptor antagonists are believed to produce anti-nausea effects through vagal afferents in the stomach, and they have been observed to be helpful in treating the acute stages of chemotherapy and radiation therapy-induced emesis.

Although the anti-diarrheal pathway action of 5HT3 antagonists remains to be determined, there are suggestions that these antagonists likely inhibit 5HT3 receptors located on intrinsic and extrinsic afferent nerve fibers in the mucosa and neurons that contribute to fast EPSPs that interact with 5-HT, all of which, when put together, would alleviate diarrhea by decreasing propulsive motility and secretion locally within the gut.

Contrasting with the anti-diarrheal therapy that 5HT3 has to offer, it has also been used to treat constipation. The benefit is availed by exploiting its potential to induce propelling motion and secretory actions in the intestine, since it desensitizes quickly, the emphasis is mostly on partial agonists [68].

14.3 5HT4

5HT4 serotonin receptors is one of the three major groups of receptors including 5HT4, 5HT6 and 5HT7 which are coupled with the G protein. They are found in the whole body profoundly seen in the GIT smooth muscles and neurons of the nervous system of the intestine. That's why also known as the "second brain of the gut". Both 5HT3 and 5HT4 receptors are also seen in the enterochromaffin cells (EC), myenteric neurons within the plexuses of submucosa of the enteric nervous system (ENS), sensory neurons seen intrinsically and extrinsically [62].

A single gene is the source of a functioning 5-HT4 receptor protein. There are different isoforms due to alternative splicing of their genes, leading to variations in their structure and function. These isoforms can have unique roles in various physiological processes and may respond differently to ligands or signaling pathways [69].

5HT4 receptor excitation by the serotonin plays crucial role in the GIT for gut motility regulation and facilitating release of neurotransmitters- acetylcholine and tachykinins, which increases gut motility in reflex and relaxes gut smooth muscles at certain areas. 5-HT4 receptors contribute physiologically for the regulation of propulsive motility along with promotion of normal motility through trophic actions. In the muscularis externa, 5-HT4 synergists act on pre-synaptic nerve terminals, that augment the production of of acetylcholine which result reflex of naturally occurring activity rather generation of neurotransmission [70]. It has been seen on infusion 5-HT4 agonists into the lumen or surface application to the mucosa that increase propulsive motility in peristaltic/ motility reflex in ex vivo motility assays. That suggests 5-HT4 receptors excessive number in the mucosa [71].

Further, 5-HT4 receptor has been demonstrated in the colonic epithelial cells which releases 5-HT on activation and also expresses the discharge of mucus from the goblet cells. Augmentation of the peristaltic reflex pathways is seen by pre-synaptic 5HT4 receptors activation on nerve ends to increase the acetylcholine production. Fluorescent evaluation of the promoter of the 5-HT4 gene in a strain of mice evaluated. That shows increased fluorescent protein, green in color which reveals that essentially all the colonic cells of rat and human expresses 5-HT4 receptor. Furthermore, 5-HT4 agonists activated 5-HT release from EC cells can be blocked by 5-HT4 receptor antagonists on mucosal application. That can also affect Cl- secretion by intestinal cells and mucus flow by goblet cells. For constipation therapy 5-HT4 receptors in epithelium could be pursued as a secure and applicable measure [72].

14.4 5HT7

5-HT7 is a member of the 5-HT4, 5-HT6, and 5-HT7 receptor subfamily, which is one of the three main classes of serotonin receptors that fall under the G protein-coupled receptor subfamily. They are presented in ileum, colon and stomach with limited presentation in the liver, spleen and kidneys and the CNS [68, 73]. It functions to relax human colonic smooth muscle, may inhibit peristalsis, and influence neuronal signaling of the abdominal pain to the central nervous system and peripheral tissues. Inhibiting 5-HT7 receptors has been theorized to exacerbate colitis presumably through elevating the degree of threshold pressure inducing intestinal peristalsis and lowering intestinal wall compliance. There are currently no clinically available selective 5-HT7 receptor ligands, however selective antagonists suited for in vivo administration are being researched [64, 68, 74, 75].

15. Serotonin and gut disorders

Although there are many disorders associated with serotonin deficiency, but here we sticking to the Inflammatory Bowel Syndrome abbreviated as IBS to stay within the scope of our discussion related to serotonin and colon at this point. IBS is a chronic gastrointestinal condition that affects 9–23% of the world's population. It is characterized by a wide range of symptoms, including discomfort in the stomach and changes in bowel movement pattern, although there is no structural damage to the gut involves as per known, thus, it is not to be confused with Inflammatory Bowel Disease (IBD) which presents with damage to the anatomical structures within the gut [71]. Based on the symptoms presenting i.e. diarrhea, constipation, both, or neither IBS is classified into four types: IBS-D, IBS-C, IBS-M, and IBS-U. IBS-D is characterized by diarrhea, IBS-C by constipation, and the third by mixed symptoms of both constipation and diarrhea. While IBS-U stays undefined [64].

One way serotonin might correlate with IBS is, Intestinal serotonin deficiency is found to lead to the weakening of the intestinal lining accompanied with constipation and increased serotonin level within the gut which is presumably promoted by the reduced SERT expression in IBS patients. SERT-P polymorphism has been found to be correlated with IBS-C subtypes in Indian populations. Abnormal serotonergic functions are also caused by prolonged shunting of tryptophan in pathway of kynurenine [66, 76].

Although mucosal 5-HT and TpH1 mRNA (which are linked to serotonin synthesis) are produced at lower levels in IBS-D patients, the gut's basic and enhanced release of serotonin (5HT) remains unchanged. This runs counter to the elevated serotonin levels shown in blood tests without platelets from people with IBS-D following meals [62]. Therapeutic agents that concentrate on the regulation of 5-HT activity in IBS include agonists and antagonists of 5-HT4 and 5-HT3 respectively. Because they minimize visceral sensitivity, 5-HT3 receptor antagonists have been used to treat diarrhea and abdominal pain, which are frequently encountered in IBS. This is very likely given that they act at 5-HT3 receptors on intrinsic neurons, which stimulate propulsive motility, and extrinsic sensory neurons, which signal pain and discomfort, unfortunately, these benefits are availed on the expense of severe side effects, thus the most of the antagonists (such as aldosterone) are tightly regulated. Certain antagonists, for instance Ondansetron, facilitate in reducing the frequency, urgency, transit duration, and consistency of stools, but the efficacy is reduced through the progression of disease. Agonists of this particular receptor is used to treat constipation predominant IBS [74]. The 5HT3R and 5HT4 antagonists repressed engine action of colon, while the 5HTR2B antagonists significantly affected colonic engine movement. 5-HT2B/C-selective antagonists SB206553, methysergide, and ritanserin impeded the proliferation of neurons in vitro by serotonin [75, 77].

5-HT4 agonists, like teaser, have proven to be both efficacious and safe in IBS presented with constipation; both acute and chronic. Unlike some other medications, they do not stimulate pain-sensing nerves or initiate muscle contractions in the gut directly. Instead, they work by leveraging natural stimuli to activate the body's reflexes. By elevating the production of certain chemical messengers that aid in gut movement (prokinetic pathways), these medications strengthen and support the natural digestive processes [78, 79].

5HT7 receptors regulates nociception, smooth muscle relaxation and is believed to have a role in the pathogenic mechanisms that underlie the visceral paresthesia correlated with IBS, which adds to the rationale to its therapeutic potential. Elevated 5HT7 expression has been observed in the hippocampus, hypothalamus, and intestine (ileum and colon) of IBS groups of rodent models as compared to controls, which might explain comorbid depression and anxiety disorders in IBS patients, as 5HT7 receptors have previously been associated with depression (**Table 4**) [80].

5-HTR	Location	Main function	Potential effects	Pathological implications	Role in therapeutics
5-HT1A	hippocampus, medial septum, amygdala, hypothalamus, cortical pyramidal neurons, and interneurons of prefrontal cortex	Inhibitory	cognitive function, mood and emotional states.	Anxiety disorders, bipolar disorder, schizophrenia, and severe depressive disorder	Frequent target for manage depression, Schizophrenia, bipolar disorder and anxiety.
5-HT1B	basal ganglia, striatum, frontal cortex and Cranial Blood Vessel	Inhibits neurotransmitter release	mood, memory, Smooth muscle contraction. Causes both vasoconstriction and vasodilation	Hypertension, migraine and thrombotic events.	5-HT1B agonists (triptans), in the pharmacotherapy of migraines
5-HT1D	Presynaptic Neuron	Inhibits neurotransmitter release	Inhibitory, Vasoconstriction Constriction of intracranial blood vessel smooth muscle	Hypertension, migraine	Same agonists as 5HT1B
5-HT2A	smooth muscle cells of blood vessel, Platelet	Regulating vascular tone and blood flow	Platelet Aggregation, Contraction	cardiovascular diseases, atheroscelerosis	Antagonists are used as antihypertensive and anticoagulants. Agonists are being explored to treat mental disorders and drug abuse.
5-HT2B	the GIT, liver, kidney, and heart	peristalsis and secretions production	Contraction	Disruption might affect neuronal differentiation	-HT2B-selective antagonists impedes differentiation of neurons in vitro by 5-HT, affects colonic engine movement.
5-HT3	Peripheral neuron, CNS	↑ transmitter release, ↑secretion	Gut motility and visceral sensitivity	IBS	Both antagonists and agonists are used in management of IBS
5-HT4	Enteric neurons, smooth muscles of GIT	↑Transmitter release, ↑ secretion	Relaxation, regular and propulsive motility	IBS	Agonists show more promising management of IBS-C/M
5-HT5A	hippocampus and neocortex	Not fully understood	learning, memory, and mood regulation.	Schizophrenia, unipolar depression and other psychiatric disorders	Therapies targeted to this receptor might show promising effects in improving memory and blood flow.

5-HTR	Location	Main function	Potential effects	Pathological implications	Role in therapeutics
5-HT6	cortex and hippocampus	Neuronal wiring and brain development	cognition, learning, memory, mood	Alzheimer's disease (AD), anxiety and cognitive decline	target for treating cognitive decline, antagonists might help in the reversal of age and pharmacological related impairment, repair of brain damage and neuronal plasticity.
5-HT7	Vascular and nonvascular smooth muscle like	Neural signaling of pain to peripheral tissues	Relaxation, immune activation & inflammatory response.	Visceral paresthesia, co-morbid anxiety and depression in IBS, may aggravate colitis	Potential target to address nociception and comorbid anxiety in IBS

Table 4.
Location, main function, potential effects and pathological implications of 5HTR.

16. Conclusion and future direction

There is a plethora of literatures available regarding serotonin and its role in the gut brain axis, despite that many questions are still left unanswered that warrants future research. There is no clinical use deduced for 5HT2B receptors and the research still remains in its infancy. 5HT7 receptors presents promising therapeutic potential since it has been found to affect both gut and brain, development of drugs targeting this specific receptor may prove to be effective to simultaneously treat the gut-induced brain disorders and vice versa, possibly without misbalancing the gut-brain equilibrium a seen with SSRIs and SNRIs. Apart from that, role of Neuronal 5HT and its synthesis by TPH2 is yet another area to be explored to investigate the complexity of ENS in relation to CNS. In conclusion, the intricate network of serotonin receptors within the bloodstream constitutes a complex interplay of 5-HTR2A, 5-HTR1B, and 5-HTR1D, guide's complicated processes involving platelet aggregation, vascular tone, and endothelial function. The vasoconstrictive effects of 5-HTR1B and 5-HTR2A expressed on smooth muscle cells are crucial for hemostasis and wound repair, while 5-HT1D receptors modulate serotonin release, impacting overall vascular constriction. Dysregulation of these receptors is implicated in migraine, hypertension, and thrombosis. Notably, the therapeutic benefits of these receptors is evident in development of triptans, designed to alleviate migraines by balancing vasoconstrictive and vasodilatory responses. Additionally, NO-releasing triptans and antagonists like ketanserin offer novel approaches to mitigate platelet activation and promote vascular health. As our understanding of these receptors deepens, future directions should focus on unraveling their intricate mechanisms, elucidating their roles in various diseases, and advancing targeted therapeutic strategies. This evolving knowledge holds the promise of shedding light on new avenues for maintaining vascular equilibrium and addressing a range of cardiovascular disorders.

Author details

Javeria Tanveer[1], Ammarah Baig[2], Rukhsana Rubeen[3], Shahana Rasheed Qureshi[4], Nosheen Bashir[1], Kanza Khan[5] and Kaneez Fatima-Shad[6]*

1 Jinnah University for Women, Karachi, Pakistan

2 Aga Khan University, Karachi, Pakistan

3 Dow University of Health Sciences, Karachi, Pakistan

4 Department of Biochemistry, Jinnah University for Women, Karachi, Pakistan

5 Department of Biotechnology, Jinnah University for Women, Karachi, Pakistan

6 Neuroscience, UTS Sydney, Sydney, Australia

*Address all correspondence to: ftmshad@gmail.com

References

[1] Kanova M, Kohout P. Serotonin—Its synthesis and roles in the healthy and the critically ill. International Journal of Molecular Sciences. 2021;**22**(9):4837

[2] Wu H, Denna TH, Storkersen JN, Gerriets VA. Beyond a neurotransmitter: The role of serotonin in inflammation and immunity. Pharmacological Research. 2019;**140**:100-114

[3] David D, Gardier A. Les bases de pharmacologie fondamentale du système sérotoninergique: application à la réponse antidépressive. L'Encéphale. 2016;**42**(3):255-263

[4] Smith C, Smith M, Cunningham R, Davis S. Recent advances in Antiemetics: New formulations of 5-HT: 3: Receptor antagonists in adults. Cancer Nursing. 2020;**43**(4):E217-EE28

[5] Conde K, Fang S, Xu Y. Unraveling the serotonin saga: From discovery to weight regulation and beyond-a comprehensive scientific review. Cell & Bioscience. 2023;**13**(1):1-11

[6] Coleman JA, Yang D, Zhao Z, Wen P-C, Yoshioka C, Tajkhorshid E, et al. Serotonin transporter–ibogaine complexes illuminate mechanisms of inhibition and transport. Nature. 2019;**569**(7754):141-145

[7] Kitson SL. 5-hydroxytryptamine (5-HT) receptor ligands. Current pharmaceutical design. 2007;**13**(25):2621-2637

[8] De Deurwaerdère P, Di Giovanni G. Serotonergic modulation of the activity of mesencephalic dopaminergic systems: Therapeutic implications. Progress in Neurobiology. 2017;**151**:175-236

[9] Seyedabadi M, Fakhfouri G, Ramezani V, Mehr SE, Rahimian R. The role of serotonin in memory: Interactions with neurotransmitters and downstream signaling. Experimental brain research. 2014;**232**:723-738

[10] Svensson E, Apergis-Schoute J, Burnstock G, Nusbaum MP, Parker D, SchiöthHB. General principles of neuronal co-transmission: Insights from multiple model systems. Frontiers in neural circuits. 2019;**12**:117

[11] Gershon M. Serotonin receptors and transporters—Roles in normal and abnormal gastrointestinal motility. Alimentary pharmacology & therapeutics. 2004;**20**:3-14

[12] Fiorica-Howells E, Maroteaux L, Gershon MD. Serotonin and the 5-HT2B receptor in the development of enteric neurons. Journal of Neuroscience. 2000;**20**(1):294-305

[13] Pergolizzi S, Rizzo G, Favaloro A, Alesci A, Pallio S, Melita G, et al. Expression of VAChT and 5-HT in ulcerative colitis dendritic cells. Acta histochemica. 2021;**123**(4):151715

[14] Ruat M, Traiffort E, Leurs R, Tardivel-Lacombe J, Diaz J, Arrang J-M, et al. Molecular cloning, characterization, and localization of a high-affinity serotonin receptor (5-HT7) activating cAMP formation. National Academy of Sciences of the United States of America. 1993;**90**(18):8547-8551

[15] Chagraoui A, Thibaut F, Skiba M, Thuillez C, Bourin M. 5-HT2C receptors in psychiatric disorders: A review. Progress in Neuro-Psychopharmacology & Biological Psychiatry. 2016;**66**:120-135

[16] Hjorth S, Bengtsson HJ, Milano S. Raphe 5-HT1A autoreceptors, but not

postsynaptic 5-HT1A receptors or β-adrenoceptors, restrain the citalopram-induced increase in extracellular 5-hydroxytryptamine in vivo. European journal of pharmacology. 1996;**316**(1):43-47

[17] Frey BN, Rosa-Neto P, Lubarsky S, Diksic M. Correlation between serotonin synthesis and 5-HT1A receptor binding in the living human brain: A combined α-[11C] MT and [18F] MPPF positron emission tomography study. NeuroImage. 2008;**42**(2):850-857

[18] Cryan JF, Valentino RJ, Lucki I. Assessing substrates underlying the behavioral effects of antidepressants using the modified rat forced swimming test. Neuroscience & Biobehavioral Reviews. 2005;**29**(4-5):547-569

[19] Doly S, Fischer J, Brisorgueil MJ, Vergé D, Conrath M. 5-HT5A receptor localization in the rat spinal cord suggests a role in nociception and control of pelvic floor musculature. Journal of Comparative Neurology. 2004;**476**(4):316-329

[20] Costa-Mallen P, Checkoway H, Fishel M, Cohen AW, Smith-Weller T, Franklin GM, et al. The EcoRV genetic polymorphism of human monoamine oxidase type a is not associated with Parkinson's disease and does not modify the effect of smoking on Parkinson's disease. Neuroscience Letters. 2000;**278**(1-2):33-36

[21] Dorszewska J, Prendecki M, Oczkowska A, Rozycka A, Lianeri M, Kozubski W. Polymorphism of the COMT, MAO, DAT, NET and 5-HTT genes, and biogenic amines in Parkinson's disease. Current Genomics. 2013;**14**(8):518-533

[22] Horjales-Araujo E, Demontis D, Lund EK, Vase L, Finnerup NB, Børglum AD, et al. Emotional modulation of muscle pain is associated with polymorphisms in the serotonin transporter gene. PAIN®. 2013;**154**(8): 1469-1476

[23] Vilaró M, Cortés R, Mengod G, Hoyer D. Distribution of 5-HT receptors in the central nervous system: An update. In: Handbook of Behavioral Neuroscience. Vol. 31. Australia NZ: Elsevier; 2020. pp. 121-146

[24] Garcia-Alloza M, Hirst W, Chen C, Lasheras B, Francis P, Ramirez M. Differential involvement of 5-HT1B/1D and 5-HT6 receptors in cognitive and non-cognitive symptoms in Alzheimer's disease. Neuropsychopharmacology. 2004;**29**(2):410-416

[25] Benhamú B, Martín-Fontecha M, Vázquez-Villa H, Pardo L, López-Rodríguez ML. Serotonin 5-HT6 receptor antagonists for the treatment of cognitive deficiency in Alzheimer's disease. Journal of Medicinal Chemistry. 2014;**57**(17):7160-7181

[26] Matthys A, Haegeman G, Van Craenenbroeck K, Vanhoenacker P. Role of the 5-HT7 receptor in the central nervous system: From current status to future perspectives. Molecular Neurobiology. 2011;**43**:228-253

[27] Vanhoutte PM. Serotonin: A forgotten signal from the blood. In: Handbook of Behavioral Neuroscience. Vol. 31. Australia NZ: Elsevier; 2020. pp. 393-409

[28] Moerland M, Kemme M, Dijkmans A, Bergougnan L, Burggraaf J. Modulation of vasoactivity and platelet aggregation by selective 5-HT receptor antagonism in humans. Journal of Cardiovascular Pharmacology. 2011;**58**(6):575-580

[29] Hu L, Wang B, Zhang Y. Serotonin 5-HT6 receptors affect cognition in a mouse model of Alzheimer's disease by regulating cilia function. Alzheimer's research & therapy. 2017;**9**(1):1-17

[30] Yamazaki M, Okabe M, Yamamoto N, Yarimizu J, Harada K, Yamazaki M. Novel 5-HT5A Receptor Antagonists Ameliorate Scopolamine-Induced Working Memory Deficit. USA: Oxford Academic; 2015

[31] Gellynck E, Heyninck K, Andressen KW, Haegeman G, Levy FO, Vanhoenacker P, et al. The serotonin 5-HT 7 receptors: Two decades of research. Experimental brain research. 2013;**230**:555-568

[32] Tabima DM, Frizzell S, Gladwin MT. Reactive oxygen and nitrogen species in pulmonary hypertension. Free Radical Biology and Medicine. 2012;**52**(9):1970-1986

[33] Thomas DR. 5-ht5A receptors as a therapeutic target. Pharmacology & therapeutics. 2006;**111**(3):707-714

[34] Sagi Y, Medrihan L, George K, Barney M, McCabe KA, Greengard P. Emergence of 5-HT5A signaling in parvalbumin neurons mediates delayed antidepressant action. Molecular Psychiatry. 2020;**25**(6):1191-1201

[35] Ramírez MJ. 5-HT 6 receptors and Alzheimer's disease. Alzheimer's research & therapy. 2013;**5**:1-8

[36] Richerson GB. Serotonin: The anti-SuddenDeathAmine? Serotonin & SUDEP. Epilepsy currents. 2013;**13**(5):241-244

[37] Jeske NA, Berg KA, Cousins JC, Ferro ES, Clarke WP, Glucksman MJ, et al. Modulation of bradykinin signaling by EP24. 15 and EP24. 16 in cultured trigeminal ganglia. Journal of Neurochemistry. 2006;**97**(1):13-21

[38] Bédard C, Wallman M-J, Pourcher E, Gould PV, Parent A, Parent M. Serotonin and dopamine striatal innervation in Parkinson's disease and Huntington's chorea. Parkinsonism & related disorders. 2011;**17**(8):593-598

[39] Blattner KM, Canney DJ, Pippin DA, Blass BE. Pharmacology and therapeutic potential of the 5-HT7 receptor. ACS Chemical Neuroscience. 2018;**10**(1):89-119

[40] Watts SW, Morrison SF, Davis RP, Barman SM. Serotonin and blood pressure regulation. Pharmacological Reviews. 2012;**64**(2):359-388

[41] Gamoh S, Hisa H, Yamamoto R. 5-hydroxytryptamine receptors as targets for drug therapies of vascular-related diseases. Biological and pharmaceutical bulletin. 2013;**36**(9):1410-1415

[42] Zhang G, Stackman RW. The role of serotonin 5-HT2A receptors in memory and cognition. Frontiers in Pharmacology. 2015;**6**:1-17

[43] Longmore J, Maguire J, MacLeod A, Street L, Schofield W, Hill R. Comparison of the vasoconstrictor effects of the selective 5-HT1D-receptor agonist L-775,606 with the mixed 5-HT1B/1D-receptor agonist sumatriptan and 5-HT in human isolated coronary artery. British journal of clinical pharmacology. 2000;**49**(2):126-131

[44] Pletscher A. The 5-hydroxy-tryptamine system of blood platelets: Physiology and pathophysiology. International Journal of Cardiology. 1987;**14**(2):177-188

[45] Nishihira K, Yamashita A, Tanaka N, Moriguchi-Goto S, Imamura T, Ishida T,

et al. Serotonin induces vasoconstriction of smooth muscle cell-rich neointima through 5-hydroxytryptamine2A receptor in rabbit femoral arteries. Journal of Thrombosis and Haemostasis. 2008;**6**(7):1207-1214

[46] Blessing WW, Seaman B. 5-Hydroxytryptamine2A receptors regulate sympathetic nerves constricting the cutaneous vascular bed in rabbits and rats. Neuroscience. 2003;**117**(4):939-948

[47] Williams MS. Platelets and depression in cardiovascular disease: A brief review of the current literature. World Journal of Psychiatry. 2012;**2**(6):114

[48] Muller CP, Cunningham KA. Handbook of the Behavioral Neurobiology of Serotonin. Australia NZ: Academic Press; 2020

[49] Tepper SJ, Rapoport AM, Sheftell FD. Mechanisms of action of the 5-HT1B/1D receptor agonists. Archives of Neurology. 2002;**59**(7):1084-1088

[50] Classey J, Bartsch T, Goadsby P. Distribution of 5-HT1B, 5-HT1D and 5-HT1F receptor expression in rat trigeminal and dorsal root ganglia neurons: Relevance to the selective anti-migraine effect of triptans. Brain Research. 2010;**1361**:76-85

[51] Blackburn TP. 5 Actions of Drugs on the Brain and CNS Disorders. Pharmacology for Chemists: Drug Discovery in Context. USA: Oxford University Press; 2017. p. 130

[52] Goadsby P, Classey J. Evidence for serotonin (5-HT) 1B, 5-HT1D and 5-HT1F receptor inhibitory effects on trigeminal neurons with craniovascular input. Neuroscience. 2003;**122**(2):491-498

[53] de Vries P. 5-Hydroxytryptamine Receptors Mediating Carotid and Systemic Haemodynamic Effects: The Relation to Acute Antimigraine Therapy. Rotterdam, The Netherlands; 1999

[54] Mendelson SD. The current status of the platelet 5-HT2A receptor in depression. Journal of Affective Disorders. 2000;**57**(1-3):13-24

[55] Moghaddam RA, Jazaeri F, Abdollahi A, Mohammadjafari R, Dehpour AR, Bakhtiarian A. Evaluation of isolated vascular response to 5HT1A, 5HT1B1D & 5HT2A Receptors Agonist & Antagonist in chronic Endotoxemic rats. Drug Research. 2019;**69**(06):352-360

[56] Schytz HW, Hargreaves R, Ashina M. Challenges in developing drugs for primary headaches. Progress in Neurobiology. 2017;**152**:70-88

[57] Tiger M, Varnäs K, Okubo Y, Lundberg J. The 5-HT(1B) receptor - a potential target for antidepressant treatment. Psychopharmacology. 2018;**235**(5):1317-1334

[58] Voronova IP, Khramova GM, Kulikova EA, Petrovskii DV, Bazovkina DV, Kulikov AV. 5-HT2A receptors control body temperature in mice during LPS-induced inflammation via regulation of NO production. Pharmacological Research. 2016;**103**:123-131

[59] Kendig DM, Grider JR. Serotonin and colonic motility. Neurogastroenterology and Motility. 2015;**27**(7):899-905

[60] Bayliss WM, Starling EH. The movements and innervation of the small intestine. The Journal of Physiology. 1901;**26**(3-4):125-138

[61] Bruta K, Vanshika BK, Bhawana. The role of serotonin and diet in the

prevalence of irritable bowel syndrome: A systematic review. Translational Medicine Communications. 2021;**6**(1):1-9

[62] Barnes NM, Neumaier JF. Neuronal 5-HT receptors and SERT. Tocris Bioscience. 2011;**34**:1-16

[63] Margolis KG, Li Z, Stevanovic K, Saurman V, Israelyan N, Anderson GM, et al. Serotonin transporter variant drives preventable gastrointestinal abnormalities in development and function. The Journal of Clinical Investigation. 2016;**126**(6):2221-2235

[64] De Ponti F. Pharmacology of serotonin: What a clinician should know. Gut. 2004;**53**(10):1520-1535

[65] Raote I, Bhattacharya A, Panicker MM. In: Chattopadhyay A, editor. Serotonin Receptors in Neurobiology. Frontiers in Neuroscience. Boca Raton (FL): CRC Press/Taylor & Francis; 2007

[66] Gardier AM, Malagié I, Trillat A, Jacquot C, Artigas F. Role of 5-HT1A autoreceptors in the mechanism of action of serotoninergic antidepressant drugs: Recent findings from in vivo microdialysis studies. Fundamental & Clinical Pharmacology. 1996;**10**:16-27

[67] Fiorica-Howells E, Maroteaux L, Gershon MD. Serotonin and the 5-HT_{2B} receptor in the development of enteric neurons. The Journal of Neuroscience. 2000;**20**(1):294-305

[68] Sbahi H, Cash BD. Chronic constipation: A review of current literature. Current Gastroenterology Reports. 2015;**17**(12):47

[69] Morita H, Mochiki E, Takahashi N, Kawamura K, Watanabe A, Sutou T, et al. Effects of 5-HT2B, 5-HT3 and 5-HT4 receptor antagonists on gastrointestinal motor activity in dogs. World Journal of Gastroenterology. 2013;**19**(39):6604-6612

[70] Terry N, Margolis KG. Serotonergic mechanisms regulating the GI tract: Experimental evidence and therapeutic relevance. Handbook of Experimental Pharmacology. 2017;**239**:319-342

[71] Takahashi K, Khwaja IG, Schreyer JR, Bulmer D, Peiris M, Terai S, et al. Post-inflammatory abdominal pain in patients with inflammatory bowel disease during remission: A comprehensive review. Crohns Colitis 360. 2021;**3**(4):otab073

[72] Leclere PG, Prins NH, Schuurkes JA, Lefebvre RA. 5-HT4 receptors located on cholinergic nerves in human colon circular muscle. Neurogastroenterology and Motility. 2005;**17**(3):366-375

[73] Brookes S, Chen N, Humenick A, Spencer NJ, Costa M. Extrinsic sensory innervation of the gut: structure and function. The Enteric Nervous System: 30 Years Later. 2016;**891**:63-69

[74] Chang W-Y, Yang Y-T, She M-P, Tu C-H, Lee T-C, Wu M-S, et al 5-HT7 receptor-dependent intestinal neurite outgrowth contributes to visceral hypersensitivity in irritable bowel syndrome. Laboratory Investigation. 2022;**102**(9):1023-1037

[75] Grider JR. Desensitization of the peristaltic reflex induced by mucosal stimulation with the selective 5-HT4 agonist tegaserod. American Journal of Physiology. Gastrointestinal and Liver Physiology. 2006;**290**(2):G319-G327

[76] Malinova TS, Dijkstra CD, de Vries HE. Serotonin: A mediator of the gut–brain axis in multiple

sclerosis. Multiple Sclerosis Journal. 2018;**24**(9):1144-1150

[77] Guzel T, Mirowska-Guzel D. The role of serotonin neurotransmission in gastrointestinal tract and pharmacotherapy. Molecules. 2022;**27**(5):1680

[78] Faure C, Patey N, Gauthier C, Brooks EM, Mawe GM. Serotonin signaling is altered in irritable bowel syndrome with diarrhea but not in functional dyspepsia in pediatric age patients. Gastroenterology. 2010;**139**(1):249-258

[79] Fayyaz M, Lackner JM. Serotonin receptor modulators in the treatment of irritable bowel syndrome. Therapeutics and Clinical Risk Management. 2008;**4**(1):41-48

[80] Kim JJ, Khan WI. 5-HT7 receptor signaling: Improved therapeutic strategy in gut disorders. Frontiers in Behavioral Neuroscience. 2014;**8**:396

Chapter 5

Serotonin and Idiopathic Scoliosis: A Review of Related Etiology and Treatment Considerations

Mark W. Morningstar

Abstract

Recent research has suggested a potential association between serotonin and idiopathic scoliosis, a complex spinal deformity of unknown origin. Studies have explored genetic associations, altered serotonin levels, and the effects of serotonin-related medications in the context of idiopathic scoliosis. Genetic studies have identified significant associations between idiopathic scoliosis and serotonin-related genes, indicating a potential genetic predisposition to the condition. Furthermore, altered serotonin levels have been observed in patients with idiopathic scoliosis, with lower serum serotonin levels reported compared to healthy controls. This chapter reviews some of the published genomic variants associated with idiopathic scoliosis. The effects of serotonin-related medications have also been investigated, highlighting potential therapeutic benefits. However, the exact mechanisms underlying the association between serotonin and idiopathic scoliosis remain unclear, warranting further research. While theoretical and animal models have shown connections between serotonin metabolism and idiopathic scoliosis, there are uncertainties when translating this information into clinical practice for primary care and other musculoskeletal specialty providers. This chapter outlines the serotonergic pathways of musculoskeletal function, serotonin clinical laboratory testing methods, as well as clinical management strategies including pharmacological, nutrient, dietary, and lifestyle-based options.

Keywords: genomics, posture, scoliosis, serotonin, spine

1. Introduction

Serotonin, also known as 5-hydroxytryptamine (5-HT), is a neurotransmitter that plays a crucial role in the central nervous system (CNS). It is derived from the amino acid tryptophan and is involved in numerous physiological functions, including mood regulation, sleep, appetite, and cognition. Structurally, serotonin is a monoamine neurotransmitter with a chemical formula of $C_{10}H_{12}N_2O$. It contains an indole ring and an amine group, making it a member of the indoleamine class of neurotransmitters. Serotonin exerts its effects by binding to specific receptors, mainly the 5-HT receptors, which are widely distributed throughout the brain and other tissues. While serotonin's involvement in mental health disorders like depression and anxiety is well-established, recent research has also highlighted its potential role in the pathogenesis of idiopathic scoliosis, a complex spinal deformity of unknown origin.

Several studies have investigated the potential link between serotonin and idiopathic scoliosis [1–11], shedding light on the role of serotonin signaling abnormalities in the pathogenesis of this complex spinal deformity. These studies have explored various aspects, including genetic associations, altered serotonin levels, and the effects of serotonin-related medications. While the exact mechanisms underlying the association between serotonin and idiopathic scoliosis are not fully understood, the findings suggest a potential involvement of serotonin in the development and progression of the condition.

One study conducted by Nelson et al. [5] examined the genetic association between idiopathic scoliosis and serotonin-related genes. The researchers analyzed a large cohort of patients with idiopathic scoliosis and identified several single nucleotide polymorphisms (SNPs) in genes related to serotonin signaling that were significantly associated with the condition. The study provided evidence for a potential genetic predisposition to idiopathic scoliosis linked to abnormalities in serotonin-related genes. Another study by Yang et al. [6] identified the MTNR1B genomic SNP in adolescent idiopathic scoliosis.

Serotonin has also been a target in animal studies of idiopathic scoliosis. In a study by Machida et al. [2], they observed the development of idiopathic scoliosis in pinealectomized chickens, as compared to a matched group that were subsequently given intraperitoneal injections of 5 hydroxytryptophan (5-HTP). They concluded that serotonin may be a therapeutic target in the treatment of idiopathic scoliosis.

In addition to genetic associations and altered serotonin levels, recent observations have been published showing the increased incidence of anxiety, mood changes, and introversion in children with idiopathic scoliosis [12]. Early interpretations of this connection were thought to be due to the psychological impacts of scoliosis treatment, such as wearing a rigid scoliosis brace [13]. However, the newer studies were conducted in children who had not participated in any treatment. Interestingly, many of these symptoms have serotonergic connections, like the serotonergic projections into the spinal and torso musculature.

While these studies provide valuable insights into the potential link between serotonin and idiopathic scoliosis, it is important to note that the underlying mechanisms are complex and multifactorial. Further research is needed to unravel the precise role of serotonin in the pathogenesis of idiopathic scoliosis and to explore potential diagnostic and therapeutic interventions targeting serotonin signaling. Nonetheless, these studies contribute to our understanding of the condition and pave the way for future investigations in this field.

2. Serotonin and the peripheral nervous system

The human body consists of a complex network of nerves that relay messages between the brain and various parts of the body. Serotonin, a neurotransmitter primarily associated with mood regulation and cognition, also plays a role in peripheral nerve pathways that terminate in the torso musculature. Understanding these serotonergic pathways is crucial for comprehending their impact on muscle function and potential therapeutic applications.

The serotonergic peripheral nerve pathways originate from the raphe nuclei, which are clusters of neurons located in the brainstem. The raphe nuclei produce serotonin and send projections that extend throughout the central nervous system (CNS) and peripheral nervous system (PNS). In the context of torso musculature, these pathways involve serotonergic fibers that innervate muscles, including the diaphragm, abdominal muscles, and back muscles.

The diaphragm, the primary muscle responsible for respiration, receives serotonergic innervation. Serotonin acts on receptors located on the diaphragmatic motor neurons, influencing their excitability and activity. Studies have shown that serotonin can modulate the motor output of the diaphragm, affecting its contraction strength and coordination [14]. Dysregulation of serotonergic pathways to the diaphragm may contribute to respiratory disorders such as sleep apnea and respiratory distress syndrome.

The abdominal muscles, including the rectus abdominis and external obliques, are involved in trunk stability and movement. Serotonergic fibers innervate these muscles, and serotonin plays a role in regulating their tone and motor control. Abnormalities in serotonergic pathways to the abdominal muscles have been implicated in conditions such as abdominal muscle weakness and spasticity.

The back muscles, encompassing the erector spinae and latissimus dorsi, are critical for posture, spinal stability, and movement. Serotonergic innervation of these muscles influences their contraction and relaxation, contributing to overall back muscle tone and coordination. Research has suggested that alterations in serotonergic signaling within the back muscles may contribute to conditions such as back pain and muscle imbalances [15].

In addition to their role in muscle function, serotonergic peripheral nerve pathways also interact with other systems in the torso. For instance, serotonin influences the gastrointestinal system, where it plays a role in regulating gastrointestinal motility and secretion. This interaction highlights the multifaceted nature of serotonergic pathways and their impact on various physiological processes.

Understanding the serotonergic peripheral nerve pathways that terminate in the human torso musculature has implications for therapeutic interventions. Modulating serotonin signaling in these pathways holds potential for managing muscle-related disorders and improving functional outcomes. Medications that target serotonin receptors, such as selective serotonin reuptake inhibitors (SSRIs), are commonly prescribed for conditions like depression and anxiety. Their effects on serotonergic pathways may extend to the modulation of torso musculature, offering therapeutic benefits for muscle-related conditions.

3. Serotonin and the central nervous system

The serotonergic central nervous pathways play a vital role in governing the body schema within the central nervous system (CNS). These pathways integrate sensory information from various sources, including cerebellar afferent input and cortical premotor inputs, to modulate motor control, body awareness, and coordination.

One crucial aspect of serotonergic pathways is their influence on the body schema, which refers to the brain's representation of the body and its position in space. Serotonin receptors are widely distributed throughout the CNS, including the cortex, basal ganglia, and cerebellum, indicating the significance of serotonin in shaping the body schema. Serotonin's effects are mediated through interactions with specific receptor subtypes, such as 5-HT1A and 5-HT2A receptors, which are abundantly expressed in key regions involved in body awareness and motor planning.

The cerebellum, a structure located at the back of the brain, receives afferent input from various sources, including proprioceptive signals from muscles and joints, as well as sensory information from the vestibular system. Serotonergic projections from the raphe nuclei provide modulatory input to the cerebellum, influencing its function and integration of sensory feedback. This integration enables the cerebellum to contribute to body schema by continuously updating motor commands based on sensory inputs.

Furthermore, the serotonergic pathways interact with cortical premotor inputs, which are responsible for motor planning and execution. These pathways involve the communication between the primary motor cortex, supplementary motor area (SMA), and the basal ganglia. Serotonin modulates the excitability of these areas, facilitating or inhibiting motor outputs based on the context and behavioral demands. By integrating cerebellar afferent input with cortical premotor inputs, serotonergic pathways contribute to the fine-tuning of motor control and coordination.

Studies have shown that dysregulation of serotonergic signaling within these pathways can lead to motor deficits and impairments in body schema. For example, disruptions in serotonin transmission have been implicated in movement disorders such as Parkinson's disease and essential tremor [16]. Serotonergic dysfunction may contribute to the motor symptoms observed in these conditions, affecting both the body schema and motor control processes.

Additionally, serotonergic pathways have implications beyond motor control. Serotonin is involved in regulating mood and emotions, with serotonin imbalances linked to psychiatric disorders such as depression and anxiety. These disorders can impact body perception and awareness, further highlighting the role of serotonergic pathways in shaping the body schema and body image.

4. Serotonin dysregulation

One potential avenue of investigation is the role of serotonin deficiency in the pathogenesis of idiopathic scoliosis. Several studies have proposed that alterations in serotonin signaling could be a contributing factor to the development of idiopathic scoliosis.

Furthermore, serotonin levels have been found to be altered in patients with idiopathic scoliosis. Winderlich and Shchekolova [17] investigated serum serotonin levels in adolescent idiopathic scoliosis (AIS) patients and found significantly higher levels compared to healthy controls. This may be due to a reduced conversion of serotonin into melatonin. The study suggested that serum serotonin may predict progressive scoliosis.

While these findings suggest a potential association between serotonin deficiency and idiopathic scoliosis, it is important to note that the relationship is complex and likely involves multiple factors. The precise mechanisms by which serotonin deficiency contributes to scoliosis development remain unclear.

Nevertheless, understanding the potential role of serotonin deficiency in idiopathic scoliosis opens up new avenues for diagnostic and therapeutic approaches. For instance, early detection of serotonin abnormalities in individuals with scoliosis may help identify those at higher risk for progression and guide treatment strategies. Moreover, medications that target serotonin signaling, such as selective serotonin reuptake inhibitors (SSRIs), which are commonly used in the treatment of depression, may hold promise for managing idiopathic scoliosis. Further research is needed to explore the efficacy and safety of such treatments.

5. Serotonin signaling dysfunction

Serotonin signaling is known to be involved in the regulation of bone growth and remodeling. Studies have shown that serotonin receptors, particularly the 5-HT2A receptor, are expressed in osteoblasts, cells responsible for bone formation. Serotonin signaling has been implicated in the balance between bone formation and resorption, and dysregulation of this signaling pathway can lead to skeletal abnormalities. In the context of idiopathic scoliosis, disruptions in serotonin signaling may contribute to an imbalance in bone growth, leading to spinal deformities.

Genetic studies have also provided evidence for the involvement of serotonin signaling dysfunction in idiopathic scoliosis. Chu et al. [11] explored the genetic factors associated with idiopathic scoliosis and found several genes involved in serotonin signaling that were significantly associated with the condition. These findings suggest that genetic variations affecting serotonin signaling pathways may contribute to the development of idiopathic scoliosis.

Furthermore, alterations in serotonin levels have been observed in patients changes in bone resorption. Maïmoun et al. [18] conducted a study that compared serum serotonin levels between individuals with adolescent idiopathic scoliosis (AIS) and healthy controls. The study found lower levels of serum serotonin were related to higher levels of bone resorption markers, suggesting a potential serotonin deficiency in these individuals. This serotonin deficiency may disrupt normal bone growth and remodeling processes, leading to the development of spinal deformities.

Animal studies have also provided insights into the role of serotonin signaling dysfunction and conversion in idiopathic scoliosis. For example, Oyama et al. [19] investigated the effects of melatonin and serotonin conversion suppression in scoliosis development in a bipedal mice model. The researchers found that forcing mice into bipedalism in the absence of melatonin (as converted from serotonin) levels resulted in abnormal spinal growth and curvature, resembling the characteristics of idiopathic scoliosis. These findings suggest that serotonin signaling dysfunction can directly impact spinal development and contribute to the pathogenesis of idiopathic scoliosis.

6. Serotonin-melatonin connection in scoliosis

Serotonin and melatonin are two closely related molecules that play important roles in various physiological processes, including sleep regulation and mood. Understanding the metabolic pathways between serotonin and melatonin sheds light on their interplay and the potential consequences of serotonin deficiency or signaling dysfunction on melatonin levels, which may have implications for the development of idiopathic scoliosis.

The synthesis of melatonin begins with the amino acid tryptophan, which is converted into 5-hydroxytryptophan (5-HTP) by the enzyme tryptophan hydroxylase. 5-HTP is then converted into serotonin by the enzyme aromatic amino acid decarboxylase. Serotonin, in turn, serves as the precursor for the synthesis of melatonin in the pineal gland. The conversion of serotonin into melatonin involves several enzymatic steps, including the actions of serotonin N-acetyltransferase (SNAT) and acetylserotonin O-methyltransferase (ASMT).

Serotonin deficiency or signaling dysfunction can lead to melatonin dysfunction through multiple mechanisms. First, reduced levels of serotonin may limit the availability of the precursor molecule for melatonin synthesis, thereby affecting melatonin production. Studies have shown that alterations in serotonin signaling, such as decreased serotonin transporter (SERT) activity, can result in lower levels of serotonin and subsequently impact melatonin synthesis.

Second, serotonin is involved in regulating the activity of enzymes responsible for melatonin synthesis. For example, SNAT, the enzyme responsible for the acetylation of serotonin to form N-acetylserotonin, is regulated by serotonin receptors. Disruptions in serotonin signaling can impair the activity of SNAT and lead to reduced melatonin synthesis. Additionally, ASMT, the enzyme responsible for the final step of melatonin synthesis, is also regulated by serotonin receptors. Altered serotonin signaling can influence ASMT activity and contribute to melatonin dysfunction.

Melatonin dysfunction, resulting from serotonin deficiency or signaling dysfunction, may have implications for the development of idiopathic scoliosis. Melatonin is involved in the regulation of bone metabolism and growth, including the regulation of osteoblast and osteoclast activity. Studies have suggested that melatonin influences the balance between bone formation and resorption, and disruptions in melatonin levels or signaling may affect skeletal development.

In idiopathic scoliosis, alterations in melatonin levels and melatonin receptor expression have been observed. For example, some studies have reported lower melatonin levels in patients with idiopathic scoliosis compared to healthy controls. Melatonin receptors, particularly MT2 receptors, have also been found to be altered in individuals with idiopathic scoliosis. These findings suggest a potential link between melatonin dysfunction and the development of spinal deformities.

The exact mechanisms by which melatonin dysfunction contributes to idiopathic scoliosis are not fully understood. However, melatonin's role in bone metabolism, its influence on osteoblast and osteoclast activity, and its potential effects on skeletal growth and development provide a plausible connection between melatonin dysfunction and spinal deformities observed in idiopathic scoliosis.

In summary, serotonin deficiency or signaling dysfunction can lead to melatonin dysfunction through various mechanisms, including reduced availability of serotonin as a precursor molecule and alterations in the activity of enzymes involved in melatonin synthesis. Melatonin dysfunction, in turn, may have implications for the development of idiopathic scoliosis due to its involvement in bone metabolism and skeletal development [20]. Further research is needed to elucidate the precise mechanisms underlying the relationship between serotonin, melatonin, and idiopathic scoliosis and explore potential therapeutic interventions targeting these pathways.

7. Diagnostic laboratory testing for serotonin

Serotonin, a neurotransmitter involved in mood regulation and various physiological processes, has garnered significant interest in clinical and research settings. Serotonin lab testing plays a crucial role in diagnosing and monitoring conditions related to serotonin imbalance. This article aims to provide an overview of the different types of serotonin lab tests available, as well as discuss their reliability and validity in assessing serotonin levels.

7.1 Types of serotonin lab tests

Blood Serotonin Level Testing: Blood tests are commonly used to measure serotonin levels. These tests involve drawing a blood sample, typically from a vein in the arm, which is then analyzed to determine the concentration of serotonin. The most commonly used method is high-performance liquid chromatography (HPLC), which separates and quantifies serotonin molecules. Blood serotonin level testing offers a relatively straightforward and accessible option for evaluating serotonin levels.

7.2 Platelet serotonin level testing

Platelets contain significant amounts of serotonin and are used as a surrogate measure of serotonin levels. Platelet serotonin level testing involves collecting a blood sample and isolating the platelets to measure their serotonin content. This testing can provide insight into serotonin uptake and storage mechanisms [20].

7.3 Urine serotonin level testing

Urinary serotonin lab testing is an important diagnostic tool used to assess serotonin levels in various clinical contexts. It involves measuring serotonin or its metabolite, 5-hydroxyindoleacetic acid (5-HIAA), in urine samples. This type of testing provides valuable information about serotonin metabolism and can be utilized in both indirect and direct forms [21].

7.4 Indirect urinary serotonin testing

Indirect urinary serotonin testing involves measuring the levels of serotonin metabolites, such as 5-HIAA, in the urine. 5-HIAA is the major metabolite of serotonin and reflects the breakdown and excretion of serotonin in the body. Indirect testing is commonly used to evaluate serotonin production and metabolism in conditions associated with serotonin overproduction, such as carcinoid syndrome and certain types of neuroendocrine tumors [22].

The most common method for indirect urinary serotonin testing is the 24-hour urine collection. During this test, the patient collects all urine produced over a 24-hour period, which provides a cumulative measure of serotonin metabolites. The collected sample is then analyzed to measure the concentration of 5-HIAA. Elevated levels of 5-HIAA may indicate increased serotonin production or impaired metabolism.

7.5 Direct urinary serotonin testing

Direct urinary serotonin testing involves measuring the actual serotonin levels in urine. Unlike indirect testing, direct testing provides information about the actual concentration of serotonin, rather than its metabolites. Direct testing is useful in assessing serotonin excretion and can be employed in research settings or specific clinical situations.

Direct urinary serotonin testing typically involves using high-performance liquid chromatography (HPLC) or similar analytical techniques to separate and quantify serotonin molecules in the urine sample. This method provides a direct measurement of serotonin levels, allowing for a more accurate assessment of serotonin status.

7.6 Interpretation and clinical considerations

Urinary serotonin testing, both indirect and direct, can offer valuable insights into serotonin metabolism and associated disorders. Elevated levels of serotonin or its metabolites, such as 5-HIAA, can indicate conditions such as carcinoid syndrome, which is characterized by excessive serotonin production by neuroendocrine tumors. Monitoring urinary serotonin levels can help assess treatment efficacy and disease progression in these cases.

It is important to consider various factors that may influence urinary serotonin levels, such as dietary intake, medications, stress, and physical activity. Certain foods, such as bananas, pineapples, and walnuts, contain serotonin precursors and may temporarily elevate urinary serotonin levels. Medications, including selective serotonin reuptake inhibitors (SSRIs) and other serotonin-modulating drugs, can also affect serotonin levels.

Furthermore, it is essential to interpret urinary serotonin test results in the context of the individual patient's clinical presentation and medical history. Diagnostic decisions should be made in conjunction with other diagnostic measures and in consultation with healthcare professionals experienced in serotonin-related disorders.

8. Reliability and validity of serotonin lab testing

Reliability and validity are essential considerations when evaluating the usefulness of serotonin lab testing.

Serotonin lab tests generally demonstrate good reliability, particularly when performed by reputable laboratories using standardized protocols. However, it is important to ensure proper sample collection, handling, and storage to maintain accuracy. Factors such as stress, medications, and diet can also influence serotonin levels, so proper preparation and standardized conditions are crucial for reliable results.

Serotonin lab tests are valid indicators of serotonin levels in the tested samples. However, it is important to note that measuring serotonin levels in peripheral samples (blood, platelets, urine) may not always reflect central serotonin levels in the brain, as serotonin cannot easily cross the blood-brain barrier. Therefore, peripheral serotonin measurements may not directly correlate with neurotransmitter activity in the brain.

Serotonin lab testing provides valuable information for assessing serotonin levels and aiding in the diagnosis and management of various serotonin-related conditions. Blood serotonin level testing, platelet serotonin level testing, and urine serotonin level testing are among the commonly employed options. These tests generally demonstrate good reliability when conducted under standardized conditions. While they are valid measures of peripheral serotonin levels, it is important to interpret results cautiously, as peripheral levels may not always reflect central serotonin activity. Clinical interpretation should consider a comprehensive assessment of symptoms, medical history, and other diagnostic measures. Consulting with healthcare professionals and specialized laboratories can provide guidance on appropriate serotonin lab testing and its application in specific clinical contexts.

9. Serotonin treatment considerations

Serotonin, a neurotransmitter that plays a crucial role in mood regulation and overall well-being, has garnered significant interest in the field of healthcare. Imbalances in serotonin levels have been associated with various mental health conditions. Various treatment options have been reported for serotonin imbalances, including pharmaceutical interventions, dietary supplements, food-based approaches, and lifestyle modifications.

9.1 Pharmaceutical options

Selective Serotonin Reuptake Inhibitors (SSRIs): SSRIs are commonly prescribed medications that work by increasing serotonin levels in the brain. Examples include fluoxetine (Prozac), sertraline (Zoloft), and escitalopram (Lexapro). SSRIs are widely used in the treatment of depression, anxiety disorders, and other mood-related conditions.

Serotonin-Norepinephrine Reuptake Inhibitors (SNRIs): SNRIs, such as venlafaxine (Effexor) and duloxetine (Cymbalta), not only increase serotonin but also affect norepinephrine levels. They are prescribed for conditions like major depressive disorder, generalized anxiety disorder, and fibromyalgia.

Tricyclic Antidepressants (TCAs): TCAs, such as amitriptyline and nortriptyline, have been used to treat depression and chronic pain conditions. While they primarily affect norepinephrine and serotonin, they have a broader impact on various neurotransmitters.

9.2 Dietary supplement options

5-Hydroxytryptophan (5-HTP): 5-HTP is an amino acid precursor to serotonin. It is commonly derived from the seeds of Griffonia simplicifolia. Supplementing with 5-HTP may increase serotonin production. However, caution should be exercised, as high doses can have adverse effects and interact with other medications.

St. John's Wort: St. John's Wort is an herbal supplement that has been used successfully to alleviate symptoms of depression [23]. Its exact mechanism of action is not fully understood, but it is believed to increase serotonin levels. It may interact with other medications [24], so consultation with a healthcare professional is essential.

9.3 Food-based options

Tryptophan-Rich Foods: Tryptophan is an essential amino acid involved in serotonin synthesis. Consuming foods rich in tryptophan, such as turkey, eggs, nuts, and seeds, may support serotonin production. However, the impact of dietary tryptophan on serotonin levels is influenced by various factors, including the presence of other amino acids in the diet.

Complex Carbohydrates: Consuming complex carbohydrates, such as whole grains, legumes, and fruits, can help regulate serotonin levels. These foods increase insulin levels, which promotes the absorption of amino acids other than tryptophan, allowing tryptophan to enter the brain more readily and support serotonin synthesis.

9.4 Lifestyle options

Lifestyle options are aimed at improving the intrinsic production of serotonin. All of the lifestyle recommendations listed below have been shown to improve serotonin levels in individuals.

Exercise: Regular physical exercise has been shown to increase serotonin levels and improve mood. Engaging in aerobic exercises like jogging, cycling, or swimming can have positive effects on serotonin synthesis and release [25].

Sunlight Exposure: Sunlight exposure stimulates the production of serotonin. Spending time outdoors, particularly in the morning or early afternoon, can enhance serotonin levels. Light therapy using special lamps may also be beneficial for individuals with seasonal affective disorder (SAD) [26].

Stress Management and Relaxation Techniques: Chronic stress can deplete serotonin levels. Implementing stress management techniques like meditation, yoga, deep breathing exercises, and mindfulness practices can help reduce stress and support serotonin balance.

Essential Oils: A study by Schneider [27] showed that 3 to 6 deep and slow inhalations using a specially designed essential oil inhaler increased urinary serotonin output and decreased cortisol output.

Since patients with a history of adolescent idiopathic scoliosis are known to exhibit neurological symptoms [12] consistent with serotonin dysfunction, [28] efforts to recommend lifestyle interventions that are easy to incorporate may promote healthier habits, as well as to improve serotonin utilization.

10. Impact of serotonin treatment on scoliosis outcomes

Although serotonin is thought to play a central role in the onset and progression of idiopathic scoliosis, few studies have attempted to evaluate the impact of serotonin imbalances on the clinical outcomes of scoliosis-specific therapy, such as physiotherapy, bracing, or surgical techniques. One such study by Morningstar et al. [29] evaluated the impact of assessing urinary neurotransmitters, including serotonin, in adolescents who had completed a short course of scoliosis-specific intensive physiotherapy. The entire cohort was divided into 2 groups: Group 1 did not pursue treatment recommendations for improving the neurotransmitter results, and Group2, who did. After 6 months, Group 2 had significantly better clinical outcomes, including Cobb angle measurements, compared to Group 1. This study demonstrates the tangible scoliosis benefits to improving serotonin levels.

11. Conclusions

In conclusion, emerging evidence suggests a potential link between serotonin and idiopathic scoliosis. Genetic associations, altered serotonin levels, and the effects of serotonin-related medications have been investigated, highlighting the involvement of serotonin signaling abnormalities in the development and progression of idiopathic scoliosis. However, further research is necessary to elucidate the underlying mechanisms and to explore the clinical implications of these findings.

Conflict of interest

The author declares no conflict of interest.

Author details

Mark W. Morningstar
ScoliSMART Clinics – Michigan, Grand Blanc, MI, USA

*Address all correspondence to: drmorningstar@scolismart.com

References

[1] Machida M. Cause of idiopathic scoliosis. Spine (Phila Pa, 1976). 1999;**24**(24):2576-2583

[2] Machida M, Miyashita Y, Murai I, Dubousset J, Yamada T, Kimura J. Role of serotonin for scoliotic deformity in pinealectomized chicken. Spine (Phila Pa 1976). 1997;**22**(12):1297-1301

[3] Dubousset J, Machida M. Le rôle possible de la glande pinéale dans la pathogénie de la scoliose idiopathique. Etudes expérimentales et cliniques [possible role of the pineal gland in the pathogenesis of idiopathic scoliosis. Experimental and clinical studies]. Bulletin de l'Académie Nationale de Médecine. 2001;**185**(3):593-602

[4] Machida M, Dubousset J, Satoh T, Murai I, Wood KB, Yamada T, et al. Pathologic mechanism of experimental scoliosis in pinealectomized chickens. Spine (Phila Pa 1976). 2001;**26**(17):E385-E391

[5] Nelson LM, Ward K, Ogilvie JW. Genetic variants in melatonin synthesis and signaling pathway are not associated with adolescent idiopathic scoliosis. Spine. 2011;**36**:37-40

[6] Yang M, Wei X, Yang W, et al. The polymorphisms of melatonin receptor 1B gene (MTNR1B) (rs4753426 and rs10830963) and susceptibility to adolescent idiopathic scoliosis: A meta-analysis. Journal of Orthopedic Science: Official Journal of the Japanese Orthopaedic Association. 2015;**20**:593-600

[7] Yang P, Liu H, Lin J, et al. The association of rs4753426 polymorphism in the melatonin receptor 1B (MTNR1B) gene and susceptibility to adolescent idiopathic scoliosis: A systematic review and meta-analysis. Pain Physician. 2015;**18**:419-431

[8] Wang H, Wu Z, Zhuang Q, et al. Association study of tryptophan hydroxylase 1 and arylalkylamine N-acetyltransferase polymorphisms with adolescent idiopathic scoliosis in Han Chinese. Spine. 2008;**33**:2199-2203

[9] Li J, Yang Z, Yu M. Association study of single nucleotide polymorphism in tryptophan hydroxylase 1 gene with adolescent idiopathic scoliosis: A meta-analysis. Medicine (Baltimore). 2021;**100**(3):e23733

[10] Pompeiano O, Manzoni D, Miele F. Pineal gland hormone and idiopathic scoliosis: Possible effect of melatonin on sleep-related postural mechanisms. Archives Italiennes de Biologie. 2002;**140**(2):129-158

[11] Chu WC, Lam WM, Ng BK, Tze-Ping L, Lee KM, Guo X, et al. Relative shortening and functional tethering of spinal cord in adolescent scoliosis - result of asynchronous neuro-osseous growth, summary of an electronic focus group debate of the IBSE. Scoliosis. 2008;**3**:8

[12] Mitsiaki I, Thirios A, Panagouli E, Bacopoulou F, Pasparakis D, Psaltopoulou T, et al. Adolescent idiopathic scoliosis and mental health disorders: A narrative review of the literature. Children (Basel). 2022;**9**(5):597

[13] Misterska E, Glowacki M, Harasymczuk J. Brace and deformity-related stress level in females with adolescent idiopathic scoliosis based on the Bad Sobernheim Stress

Questionnaires. Medical Science Monitor. 2011;**17**(2):CE83-CR90

[14] Charsar BA, Brinton MA, Locke K, Chen AY, Ghosh B, Urban MW, et al. AAV2-BDNF promotes respiratory axon plasticity and recovery of diaphragm function following spinal cord injury. The FASEB Journal. 2019;**33**(12):13775-13793

[15] Xiang HB, Liu C, Liu TT, Xiong J. Central circuits regulating the sympathetic outflow to lumbar muscles in spinally transected mice by retrograde transsynaptic transport. International Journal of Clinical and Experimental Pathology. 2014;**7**(6):2987-2997

[16] Politis M, Loane C. Serotonergic dysfunction in Parkinson's disease and its relevance to disability. Scientific World Journal. 2011;**11**:1726-1734

[17] Winderlich ME, Shchekolova NB. Predicting scoliosis progression in children and adolescents measuring serotonin blood level. Genij Ortopedii. 2023;**29**(3):299-306

[18] Maïmoun L, Guillaume S, Lefebvre P, Philibert P, Bertet H, Picot MC, et al. Is serum serotonin involved in the bone loss of young females with anorexia nervosa? Hormone and Metabolic Research. 2016;**48**(3):174-177

[19] Oyama J, Murai I, Kanazawa K, Machida M. Bipedal ambulation induces experimental scoliosis in C57BL/6J mice with reduced plasma and pineal melatonin levels. Journal of Pineal Research. 2006;**40**(3):219-224

[20] Burwell RG, Dangerfield PH, Moulton A, Anderson SI. Etiologic theories of idiopathic scoliosis: Autonomic nervous system and the leptin-sympathetic nervous system concept for the pathogenesis of adolescent idiopathic scoliosis. Studies in Health Technology and Informatics. 2008;**140**:197-207

[21] Kema IP, Schellings AM, Meiborg G, Hoppenbrouwers CJ, Muskiet FA. Influence of a serotonin- and dopamine-rich diet on platelet serotonin content and urinary excretion of biogenic amines and their metabolites. Clinical Chemistry. 1992;**38**(9):1730-1736

[22] Marc DT, Ailts JW, Campeau DC, Bull MJ, Olson KL. Neurotransmitters excreted in the urine as biomarkers of nervous system activity: Validity and clinical applicability. Neuroscience and Biobehavioral Reviews. 2011;**35**(3):635-644

[23] Ng QX, Venkatanarayanan N, Ho CY. Clinical use of Hypericum perforatum (St John's wort) in depression: A meta-analysis. Journal of Affective Disorders. 2017;**210**:211-221

[24] Hu Z, Yang X, Ho PC, Chan SY, Heng PW, Chan E, et al. Herb-drug interactions: A literature review. Drugs. 2005;**65**(9):1239-1282

[25] Da Silva SR, Galdino G. Endogenous systems involved in exercise-induced analgesia. Journal of Physiology and Pharmacology. 2018;**69**(1):3-13

[26] Pjrek E, Friedrich ME, Cambioli L, Dold M, Jäger F, Komorowski A, et al. The efficacy of light therapy in the treatment of seasonal affective disorder: A Meta-analysis of randomized controlled trials. Psychotherapy and Psychosomatics. 2020;**89**(1):17-24

[27] Schneider R. The increase in urinary serotonin and decrease in salivary cortisol concentrations following direct inhalations of concentrated essential oils is not induced by non-specific

effects. Endocrine Regulations. 2021;**55**(4):215-223

[28] Correia AS, Vale N. Tryptophan metabolism in depression: A narrative review with a focus on serotonin and kynurenine pathways. International Journal of Molecular Sciences. 2022;**23**(15):8493

[29] Morningstar MW, Siddiqui A, Dovorany B, Stitzel CJ. Can neurotransmitter status affect the results of exercise-based scoliosis treatment? Results of a controlled comparative chart review. Alternative Integrative Medicine. 2014;**3**:4

Chapter 6

Serotonin: Its Functional Role in Plants

Kiran Bala

Abstract

Serotonin, also known as 5-hydroxyamine, is an indoleamine that plays crucial roles as a neurotransmitter and hormone regulator in various physiological processes across the animal kingdom. This essential signaling molecule is synthesized from the aromatic amino acid tryptophan and is found in virtually all living organisms. Over the last few years, enormous research has been done on this biomolecule. In plants, they are found to be involved in several metabolic and developmental functions. Despite its widespread importance in plants still many things to understand about the mechanism of action of this biomolecule. Therefore, this chapter focuses on the current knowledge of the role of serotonin in plants.

Keywords: serotonin, neurotransmitters, physiological functions, phytoserotonin, plant hormones

1. Introduction

Serotonin is an important biochemical molecule found in both plants, as well as in animals. In plants, it is known as phytoserotonin. Initially, it was reported in the legume known as *Macuna pruriens* [1]. Later, it was found in almost 42 plant species from almost 20 different families [2]. A significant amount of the serotonin was reported in these plants. Chemically, it is known as 5-HT (5-hydroxytryptamine). It is actually a neurotransmitter recognized for its role in the mammalian central nervous system, now identified in all forms of the life from bacteria to higher eukaryotes [3]. In animal biology, especially in humans and vertebrates, numerous physiological roles of serotonin have been identified. In the beginning, it was known as enteramine as was discovered in the gut enterochromaffin cells [3], later named as serotonin and considered as one among the most ancient molecules originated first in prokaryotic life forms [4–7]. Many useful functions of serotonin have been identified in animal science. It is involved to regulate the anxiety, sleep, and moods in the mammals [8] also, known for its hallucinogenic drug effect [9]. But in the case of plants less is known about its functional role despite been discovered in plants shortly after its discovery in the animals. Many biosynthetic products of serotonin in plants were identified in mid-1990s [10]. Serotonin is abundant in parenchyma of vascular bundle, companion cell and xylem cell, and vascular bundle of fruit wall of the banana [11]. The concentration of serotonin is found different in various parts of plant. For instance, in leguminous plant, *Friffonia simplicifolia* lower levels of serotonin are detected in its leaves, while significant amount of serotonin is present in its seeds [12]. Mostly, they are found in fruits vegetables, and seeds [13]. Even in fruits, the distribution of serotonin is not uniform [9, 14]. The amount of serotonin get increased as the fruit get ripen [15–18].

2. Biosynthesis of serotonin

In animals, serotonin is produced by tryptophan in two-step processes by the two enzymes tryptophan hydroxylase and aromatic L-amino acid decarboxylase, (1) where tryptophan hydroxylase acts as rate-limiting step. Tryptophan is an essential amino acid, which is needed for synthesis of not only serotonin but also melatonin and auxin, [19–21]. Tryptophan first catalyzed into tryptamine by tryptophan decarboxylase. Tryptamine is catalyzed by tryptamine-5-hydroxylase to form serotonin (2). However, in certain plants, such as *Hypericum perforatum* serotonin is synthesized from the hydroxytryptophan as in mammals [22]. Hydroxylation of tryptophan leads to formation of 5-oxytryptophan in the presence of tryptophan-5-hydroxylase. Later, 5-oxytryptophan is decarboxylated by decarboxylase that gives serotonin (**Figure 1**).

3. Physiological role of serotonin in plants

A lot of studies have been carried out to find out the role of serotonin in the vertebrates, whereas interest in phytoserotonin was prevented. It could be due to the less obvious role of serotonin in plants. In animals, it was first identified in 1868, whereas, in plants, it was discovered in mid-1990s. Now, serotonin is considered as an important phytohormone and stress defense compound [23, 24]. Along with melatonin, it has many roles in plant growth and survival processes [25]. A number of studies, which show its role in shoot branching [26], flowering [27], xylem sap exudation [9], ion permeability [28], morphogenesis [29], reproduction [30], germination [31], senescence [32], protection against stress [33], and root architecture (**Figure 2**) [34]. Different roles of serotonin are organized in **Table 1**. Few of them are:

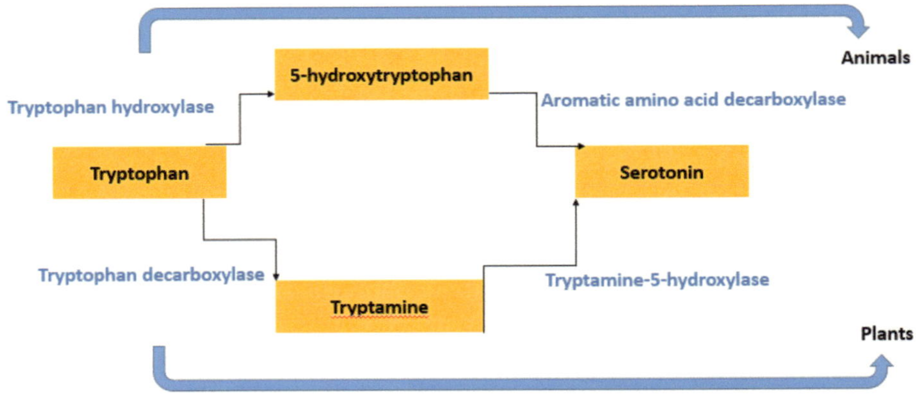

Figure 1.
Biosynthesis of serotonin.

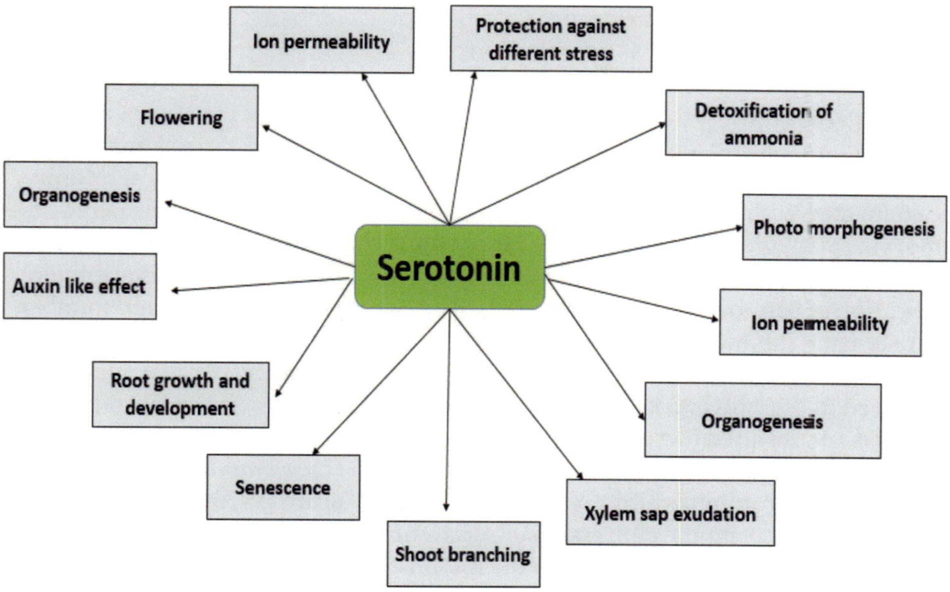

Figure 2.
Different functions of the serotonin in the plants.

S. no	Functions	Plant	References
1.	Shoot branching	Arabidopsis	[35]
2.	Flowering	Datura	[27]
3.	Ion permeability	Pea Chloroplast	[36]
4.	Xylem sap exudation	Rice	[37]
5.	Morphogenesis & growth regulator	*Mimosa pudica* L Datura	[27, 38]
6.	Root architecture	*Hypericum perforatum* L, *Arabidopsis*	[34, 39]
7.	Reproduction	Datura	[27]
8.	Germination	*Capsicum anuum* L	[31]
9.	Senescence	Rice	[32]
10.	Protection against stress	Hazelnuts	[40]
14.	Coleoptile growth	Rice, plant tissue culture, *Hypericum perforatum*	[41–43]
16.	Stress defense compounds	Arabidopsis	[44]
17.	Photosynthesis	*Chara australis*	[45]
18.	Spikelet fertility & Increased stomata conductance	Rice	[9]
19.	Signaling	Arabidopsis	[46]
20.	Detoxification of ammonia	Walnut	[47]
21.	Fruit ripening	Tomato	[48]

Table 1.
Effect of serotonin on plant growth.

3.1 Growth regulation by serotonin

Extensive studies are carried out to investigate the role of serotonin on shoot branching. As we know that many of the factors involved in the regulation of plant morphogenesis. Auxin and melatonin are among the highly recognized metabolite of tryptophan along with serotonin. Various reports on signaling role of serotonin are connected with other phytohormones [26, 49]. Serotonin is now considered an important plant growth regulator [27, 38] as studies have shown its involvement in mediating vegetative growth and morphogenesis of the plants [26]. Exogenous application of serotonin in tissue culture leads to increase in shoot size and number, whereas using inhibitors revert its impact [41, 42, 50–53]. In other studies in rice (*Oryza sativa* L) with deficiency of the enzyme serotonin N-acetyltransferase, which is involved in the conversion of serotonin to melatonin, serotonin concentration increased and resulted increased coleoptile growth [43]. An increase in endogenous serotonin level result in increased shoot production in *H.perforatum* [42], while application of serotonin inhibitors decrease shoot in culture [54, 55]. Similarly, inhibitors that prevent the conversion of serotonin to melatonin result in an increase in serotonin concentration, leading to the inhibition of auxin-induced rooting and the promotion of shoot growth [55]. In another study, application of serotonin in the presence of salt in sunflower seedlings increased primary root growth [39].

Serotonin along with melatonin is responsible for root elongation, formation and growth of lateral, and adventitious roots [56], thereby altering the root architecture. Many pathways are involved in their mode of action. They may act along with auxin or through independent pathways. As they are known for their role during stress, they interact with molecules like ROS, NO, and plant growth regulators. Examination of gene expression has shown that serotonin along with melatonin induced promotion of root induction and growth. Studies show that the accumulation of ROS in root tips regulates the formation of the primary growth. As melatonin and serotonin are also synthesized from tryptophan. Studies have shown that serotonin is responsible for root production in mimosa, walnut, sunflower, and *Arabidopsis thaliana* [56]. These have been explained by several theories. According to one theory in walnut culture, serotonin is converted into auxin, which has been known for apical dominance and root growth, which then induces rooting [57]. In another study, it was observed that serotonin in low concentration promotes lateral root primordial while in high conc. It has opposite effect on them [46]. Root growth and root hair development but adventurous root formation get enhanced. The authors, therefore, suggested that instead of enhancing auxin content, serotonin has an antagonistic effect. Serotonin treatment in sunflower increased primary root and hypocotyl length. Therefore, limited studies emphasize the need for further detailed studies on them. Serotonin elicits tissue-specific inhibitory effects on auxin-responsive genes at the site of primary and adventitious roots and also in lateral root primordial [9]. It seems that serotonin promotes root growth partially independent of auxin activity. Researchers of the Chinese Academy of Sciences compared the physiological response of Arabidopsis to exogenous melatonin and serotonin-mediated metabolism. The study has shown that moderate concentration of melatonin did not affect primary root growth but induce lateral root formation [35].

3.2 Role of serotonin in flowering

Flowers are associated with reproduction in the plants. Many factors influence its development. Rapid growth and development the oxidative environment in these tissues can lead to underdeveloped reproductive structure if anti-oxidative protection is absent [58, 59]. Along with melatonin serotonin gives protection for the growing flowers and seeds. This effect is noticed in various species such as *Datura metal, Hypericum perforatum, Malus domestica, Vitis vinifera, and Prunus avium*. These studies have shown that the serotonin concentration is higher in reproductive tissue than vegetative tissue [59–63]. Serotonin is predominantly distributed in reproductive structures. A study has revealed that *Griffonia simplicifolia* leaves have a minimal amount of serotonin, while the seeds have a high concentration [9, 64]. Serotonin gets accumulated in fruits [64, 65]. It may act as signal molecule in the reproductive development. Though the study on the activity of serotonin in plant reproduction is restricted, the existing literature suggests it has a strong impact on pollen germination *in vivo,* as well *in vitro* studies. During pollen growth in St. John's wort the shift from tetrad to uninucleate phase results increase in serotonin concentration further suggesting its role in reproductive growth [63, 66]. In rice, wheat, and soybean. Each of small molecules enhanced the germination of seeds in the presence of abscisic acid, although potent inhibitor of ABA [67].

3.3 Antioxidative properties and anti-stress properties of serotonin

As serotonin is biochemically indolamine made up of indole backbone with the sidechain ethylamine derived from the tryptophan. In the last century, numerous physiological roles of serotonin were discovered in animal science. Several studies have proposed that serotonin acts as antioxidant in the animals [67]. It regulates abiotic stress-induced plant growth inhibition possibly by modification of hormone metabolism [68–70]. Structurally, auxin is more similar to serotonin and melatonin. Serotonin is also known for gene regulation with auxin-responsive related pathways [71]. Accumulation of serotonin has been observed in the leaves of rice plants in the response of biotic stress [72]. Genes of tryptophan biosynthetic pathway show coordination with genes of serotonin and auxin biosynthesis under abiotic and biotic stress [73]. The enormous production of serotonin in senescing rice leaves, which has been identified by chlorophyll loss, lipid peroxidation of membrane, increased reactive oxygen species (ROS), and induced senescence-related genes. Serotonin concentration get increased after salt stress. In another study, $CdCl_2$ treatment inhibited the serotonin N-acetyltransferase gene, thereby maintaining high levels of serotonin. When serotonin is provided exogenously, it gives protection to *Brassica napus* L against salt stress [74]. It can alleviate the growth inhibition in seedlings of the same plant under salt stress. Under mild drought conditions, serotonin improves the yield [32, 58]. Exogenous serotonin on tomato seedling during drought and salt stresses have shown that serotonin has strong antioxidative effect [75]. Serotonin also resulted increase in the biomass and isoflavones content, cell division, ethylene, and isoflavones biosynthesis under temperature stress in soybean cell culture [76].

3.4 Role of serotonin in senescence

Accumulation of serotonin is known to play a protective role against ROS, leading to a delay in senescence [77]. Serotonin prevents the accumulation of the toxic metabolite due to its powerful antioxidant activity in the senesced leaves. Because of its antioxidative activity, serotonin protects xylem parenchyma during senescence-induced oxidative damage. Serotonin overexpression in plants shows delay in senescence in rice leaves, whereas transgenic plants with low serotonin expression show fast senescence [73]. Serotonin relieves the accumulation of harmful biomolecules tryptamine by its antioxidant activity in the senescenced leaves. Further physiological analysis indicated that exogenous serotonin alleviates iron deficiency-induced leaf chlorosis [78] and improves drought and salt tolerance in tomato seedlings [79]. Serotonin shows the slow senescence in corn leaves, via calcium signal transduction, interacts with phosphatidylinositol, and maintains the chlorophyll content [80].

3.5 Contribution of serotonin in photosynthesis

Serotonin is believed to be localized in the chloroplast. There is a lot of evidence that shows its role in the maintenance of the photosynthetic tissues [36, 81, 82]. According to one study, isolated chloroplast of pea depicted that serotonin is capable of enhancing efflux of magnesium and calcium. Serotonin also helps in mediating light sensing in plants *via* phytochrome modification activity and helping in signaling networks [9]. The chlorophyll content gets decreased due to salinity was increased by exogenous serotonin [74]. Serotonin levels increased significantly under longer wavelength treatment and in dark conditions [9]. In rice plants, serotonin was found to improve stomatal conductance, spikelet fertility, and yield under mild drought stress conditions [9].

3.6 Interaction of serotonin with phytochrome

One of the most important roles of the serotonin is that it recognizes light and regulates circadian and seasonal rhythms. Phytochrome is used in this process through which serotonin interacts, therefore, induce diverse metabolic activities. Serotonin stimulates phosophoinositide turnover, therefore, modulates the red light effect, enhances the nitrate reductase transcription, and inhibits phy-I transcript accumulation [38, 83]. Many reports are indicating interaction of serotonin with phytochrome in an important signal cascade. First, it was reported in the eighties when external application of serotonin was capable of takeoff the effect of red light. It affects phytochrome either by activating it or by modifying the signal transduction pathway. In one study, it was observed that serotonin application could mimic the calcium uptake observed in light-grown culture of the protoplast [84, 85]. Besides stimulating the effects of red light exposure in plants, endogenous serotonin levels also get decreased in response to red light exposure to yellow or green light in *Sedum morganianum* E. Walther. Serotonin usually found in high concentration during the day time and lowered during the night [71]. Biosynthesis of indoleamine and role of serotonin along with melatonin has been well characterized in yeast, bacteria, and mammals [86, 87]. Many studies show serotonin treatment mimic the effects of red light exposure in other physiological processes such as fruit ripening, senescence, and leaf abscission. Interaction between serotonin with metabolite phenylpropanoid suggests its contribution in the maintance of chlorophyll pigments such as chlorophyll, as well as anthocyanin [88, 89].

3.7 Effect upon ammonia

However, serotonin is found during seed development, how exactly it functions out there is not known yet. Many studies indicate probably it is involved in the detoxification of the ammonia, hence prevent the embryo. In drying seed, serotonin assists to remove accumulating ammonia. Ammonia is metabolized in L-tryptophan. After decarboxylation, it gives tryptamine. Hydroxylation of tryptamine by cytochrome

P450 monooxygenase forms serotonin [47]. Similar results of serotonin accumulation in cotyledon of walnut were observed. Here, serotonin is involved in the detoxification of ammonia reaction, thus prevent delicate plant tissue [90]. Serotonin was also detected in embryos of *Juglans mandshurica* histochemically [37, 91]. Further studies are needed to explore detail protective pathway of serotonin.

4. Conclusion and future perspectives

Though much new information is coming out about its regulatory role in plants, it is one of the most primitive biomolecules that evolved on Earth and is considered one of the most important molecules. From studies, it is now evident that it carries out diverse functions in plants. So, there is a growing interest among plant researchers to study the effect of serotonin on various plant systems. No doubt a new branch of phytoserotonin has emerged, where lots of things need to be worked out. In humans, it is involved in many vital roles. The discovery of serotonin in plants used in the treatment of human disorders provides a new route for the investigation of medicinally active compounds. Diverse roles of serotonin in plants are identified. As molecules regulate morphogenesis, they can be used as potential modulators in tissue culture regeneration techniques. The antistress and detoxification activities of serotonin further need to be investigated in detail. Moreover, auxin and serotonin biosynthesis are associated with tryptophan. Tryptophan is an important precursor for various metabolites. Thus, it is possible that both auxin and serotonin are related. So, it is important to find out the possibility of auxin-serotonin crosstalk or the crosstalk of serotonin with other hormones during different regulatory pathways.

Author details

Kiran Bala
Swami Shraddhanand College, University of Delhi, Delhi, India

*Address all correspondence to: kiranbalain@yahoo.com

References

[1] Veenstra-Vander Weele J, Anderson GM, Cook EH. Pharmacogenetics and serotonin system; initial studies and future directions. European Journal of Pharmacy. 2000;**410**:165-181. DOI: 10.1016/s0014-2999(00)00814-1

[2] Roshchina VV. Neurotransmitters in Plant Life. Enfield: Science Publishers; 2001. pp. 4-81. DOI: 10.1093/aob/mcg113

[3] Vially M, Erspamer V. Ricerche sul secreto delle cellule enterocromomaffini. Zeitschrift fur Zellforschung and Mikroskopische Anatomie. 1937;**27**: 81-99. DOI: 10.1007/BF00391792

[4] Levy RJ. Serotonin transporter mechanism and cardiac disease. Circulation. 2006;**113**:2-4. DOI: 10.1161/ CIRCULATIONAHA. 105.593459

[5] Izzati-Zade KF. The role of serotonin in the pathogenesis and clinical presentations of migraine attacks. Neuroscience and Behavioral Physiology. 2008;**38**:501-505. DOI: 10.1007/ s11055-008-9008-3

[6] Olivier B. Serotonin: A never ending story. European Journal of Pharmacology. 2015;**753**:2-18. DOI: 10.1016/j.ejphar.2014.10.031

[7] Yasin T. Parkinson's disease, subthalamic nucleus and serotonin. Frontiers in Neuroscience. 2009;**3**. DOI: 10.3389/conf.neuro.01.2009.16.119

[8] JB-Cazenave, Bhartiya R, Barriere G, Delbecque JP, Bouguiyound N, Giovanni GD, et al. Serotonin in animal cognition and behavior. International Journal of Molecular Sciences. 2020;**21**(5):1649. DOI: 10.3390/ ijms21051649

[9] Ramakrishna A, Giridhar P, Ravishanker GA. Phytoserotonin: A review. Plant Signaling & Behavior. 2011;**6**(6):800-809. DOI: 10.4161/ psb.6.6.15242

[10] Dubbels R, Reiter RJ, Klenke E, Goebel A, Schnakenberg E, Ehlers C, et al. Melatonin in edible plants identified by radioimmunoassay and by high performance liquid chromatography mass spectrometry. Journal of Pineal Research. 1995;**18**:28-31. DOI: 10.1111/j.1600-079x.1995.tb00136.x

[11] Kimura M. Fluorescence histochemical study on serotonin and catecholamine in some plants. Japan Journal of Pharms. 1968;**18**:162-168. DOI: 10.1254/jjp.18.162

[12] Fellows LE, Bell EA. Indole metabolism in Piptadenia peregrine. Phytochemistry. 1971;**10**:2083-2091

[13] Kema IP, de Vries EG, Muskiet FAJ. Clinical chemistry of serotonin and metabolites. Journal of Chromatography B. 2000;**747**:33-48. DOI: 10.1016/s0378-4346(00)00341-8

[14] Udenfriend S, Weissbach H, Clark CT. The estimation of 5-hydroxytryptamine (serotonin) in biological tissue. The Journal of Biological Chemistry. 1955;**215**:337-344

[15] Foyratt JR, Parratt JM. 5-hydroxytryptamine in pineapples. Journal of Pharmacy and Pharmaceutical. 1961;**13**:382-383. DOI: 10.1111/j.2042-7158.1961.tb11840.x

[16] Erspamer V. Recent research in the field of 5-hydroxy tryptamine and related indoealkylamins. In: Jucker E, editor. Progress in Drug Research. Basel,

Stuttgart: Birkhaser Verlag; 1961. pp. 151-367. DOI: 10.1007/978-3-0348-7041-2_3

[17] Bergmann L, Grosse W, Ruppel HG. Serotonin in *Juglans regia*. Planta. 1970;**94**:47-59. DOI: 10.1007/BF00386608

[18] Nicasio MDP, Villrreal ML, Gillet F, Bensaddek L, Fliniaux MA. Variation in the accumulation levels of N,N-Dimethyltryptamine in micropropagated trees and in vitro culture of Mimosa Tenuiflore. Natural Product Research. 2005;**19**:61-67. DOI: 10.1080/14786410410001658860

[19] Pelagio-Flores R, Lopez-Bucio J. In: Ravishanker GA, Ramakrishna A, editors. Serotonin and Melatonin Their Functional Role in Plants, Food Phytomedicine and Human Health. Boca Raton, FL, USA: CRC Press; 2016. pp. 97-110. DOI: 10.1201/97813

[20] Erland LAE, Saxena PK. Mammalian neurotransmitter are important signals mediating plant morphogenesis. In: Sopory S, editor. Sensory Biology of Plants. Singapore: Springer; 2019. pp. 411-449. DOI: 10.1007/978-981-13-8922-1_16

[21] Schroder P, Abele C, Gohr P, Stuhlfauth-Roisch U, Grosse W. Latest on the enzymology of serotonin biosynthesis in walnut seeds. Advances in Experimental Medicine and Biology. 1999;**467**:637-644. DOI: 10.1007/978-1-4615-4709-9_81

[22] Murch SJ, Krishna Raj S, Saxena PK. Tryptophan is a precursor for melatonin and serotonin biosynthesis in vitro regenerated St. John. Wort (*Hypericum perforatum* L cv Anthos) plants. Plant Cell Reports. 2000;**19**:698-704

[23] Kang K, Kang S, LeeK PM, Back K. Enzymatic and serotonin biosynthesis in plants. Plant Signaling & Behavior. 2008;**3**:289-390. DOI: 10.1007/s002990000206

[24] Bhowal B, Bhattacharjee A, Goswami K, Mishra NS, Pareek SLS, Kour C, et al. Serotonin and melatonin biosynthesis in plants: Genome-Wide identification of the genes and their expression reveal a conserved role in stress and development. International Journal of Molecular Sciences. 2021;**22**(20):11034. DOI: 10.3390/ijms222011034

[25] Hattori A, Migitaka H, Ligo M, Itoh M, Yamamoto K, Ohtani Kaneko R, et al. Identification of melatonin in plants and its effects on plasma melatonin levels and binding to melatonin receptors in vertebrates. Biochemical and Molecular Biology. 1995;**35**:637-634

[26] Erland LAE, Murch SJ, Reiter RJ, Saxena PK. A new balancing act: The many roles of melatonin and serotonin in plant growth and development. Plant Signaling & Behavior. 2015;**10**(11):e1096469. DOI: 10.1080/15592324.2015.1096469

[27] Murch SJ, Alan AR, Cao J, Saxena PK. Melatonin and serotonin in flowers and fruits of Datura metal L. Journal of Pineal Research. 2009;**47**(3):277-283. DOI: 10.1111/j.1600-079X.2009.00711.x

[28] Akula R, Mukherjee S. New insights on neurotransmitters signaling mechanism in plants. Plant Signaling & Behavior. 2020;**15**:6. DOI: 10.1080/15592324.2020.1737450

[29] Erland LAE, Shukla MR, Singh AS, Murch SJ, Saxena PK. Melatonin and serotonin; mediators in the symphony of plant morphogenesis. Journal of Pineal Research. 2018;**64**:2. DOI: 10.1111/jpi.12452

[30] Erland LAE, Turi CE, Saxena P. Chapter-2- in plants: Origin, functions and implications. Serotonin. 2019:23-46. DOI: 10.1016/B978-0-12-800050-2.00002

[31] Araz O, Yildirim E, Ekinci M. Alleviation of the germination inhibitory effect of salt stress in pepper (*Capsicum anuum* L) seeds by serotonin. Journal of Agricultural production. 2021;**2**(2):48-54. DOI: 10.29329/agripro.2021.360.2

[32] Kang K, Kim YS, Park S, Back K. Senescence induced serotonin biosynthesis and its role in delaying senescence in rice leaves. Plant Physiology. 2009;**150**(3):1380-1393. DOI: 10.1104/pp.109.138552

[33] Kaur H, Mukerjee S, Balsuka F, Bhatia SC. Regulatory roles of serotonin and melatonin in abiotic stress tolerance in plants. Plant Signal Behaviour. 2015;**10**:11. DOI: 10.1080/15592324.2015.1049788

[34] Murch SJ, Campbell SSB, Saxena PK. The role of serotonin and melatonin in plant morphogenesis: Regulation of auxin induced root organogenesis in cultured explants of St. John's wort (*Hypericum perforatum L*). In Vitro Cellular & Developmental Biology-Plant. 2009;**37**:786-793. DOI: 10.1007/s11627-001-0130

[35] Manchester LC, Tan DX, Reiter RJ, Park W, Monis K. High levels of melatonin in the seeds of edible plants: Possible function in germ tissue protection. Life Sciences. 2000;**67**:3023-3029. DOI: 10.1016/S0024-3205(00)00896

[36] Akcay UC, Okudan N. Exogenous serotonin improves drought and salt tolerance in tomato seedlings. Plant Growth Regulation. 2023;**101**:239-249. DOI: 10.1007/s10725-023-01016-x

[37] Regulla I. The presence of serotonin in the embryo of zjuglans mandshurica maxim. Acta Botanica Croatica. 1985;**44**:19-22

[38] Yabut J, Crane JD, Green AE, Keating DJ, Ihan W, Steinberg GR. Emerging roles for serotonin in regulating metabolism; New implications for an ancient molecule. Endocrine Reviews. 2019;**40**(4):1092-1107. DOI: 10.1210/er.2018-00283

[39] Pelago-Flores R. Serotonin, a tryptophan derived signal conserved in plants and animals, regulates root system architecture probably acting as a natural auxin inhibitors in *Arabidopsis thaliana*. Plant and Cell Physiology. 2011;**52**(3):490-508. DOI: 10.1093.pcp/pcr006

[40] Kiyoon K, Sei K, Kyunglin L, Munyoung P, Back K. Enzymatic features of serotonin biosynthetic enzymes and serotonin biosynthesis in plants. Plant Signaling & Behavior. 2008;**3**(6):389-390. DOI: 10.4161/psb.3.6.5401

[41] Skoog F, Miller CO. Chemical regulation of growth and organ formation in plant tissue cultured in vitro. Symposia of the Society for Experimental Biology. 1957;**11**:118-130

[42] Murch SJ, Campbell SSB, Saxena PK. The role of serotonin and melatonin in plant morphogenesis: Regulation of auxin-induced root organogenesis in vitro cultured explants of St. John's Wort (*Hypericum perforatum* L). In Vitro Cell Development Biology Plant. 2001;**37**:786-793. DOI: 10.1007/s11627-001-0130-y

[43] Byen Y, Back K. Low melatonin production by suppression of either serotonin N-acetyltransferase or N-acetylserotonin methyltransferase in rice causes seedling growth retardation with yield penalty, abiotic stress

susceptibility and enhanced coleoptile growth under anoxi. Journal of Pineal Research. 2016;**60**:348-359. DOI: 10.1111/jpi.12317

[44] Misra V, Sarkar AK. Serotonin: A frontline player in plant growth and stress responses. Physiologia Plantrum. 2023;**175**(4):e13968. DOI: 10.1111/ppl.13968

[45] Ayyanath M-M, Shukla MR, Saxena PK. Indolamines impart abiotic stress tolerance and improve reproductive traits in hazelnuts. 2023;**12**(6):1233. DOI: 10.3390/plants12061233

[46] Pelagio-Flores R, Ortiz-Castro R, Mendez-Bravo A, Macias-Rodriguez L, Lopez-Bucio J. Serotonin a tryptophan derived signal conserved in plants and animals regulates root system architecture probably acting as a natural auxin inhibitor in *Arabidospsis thaliana*. Plant & Cell Physiology. 2011;**52**:490-508. DOI: 10.1093/pcb/pcr006

[47] Erland LAE, Turi CE, Saxena PK. Serotonin: An ancient molecule and an important regulator of plant processes. Biotechnology Advances. 2016;**34**:1347-1361. DOI: 10.1016/j.biotechadv.2016.10.002

[48] Kang S, Kang K, Lee K, Back K. Characterization of tryptamine-5-hydroxylase and serotonin synthesis in plants. Plant Cell Reports. 2007;**26**:2009-2015. DOI: 10.1007/s00299-007-0405-9

[49] Haduch A, Bromek E, Kuban W, Daniel WA. The engagement of cytochrome P450 enzyme in tryptophan metabolism. Meta. 2023;**13**:5-629. DOI: 10.3390/metabo13050629

[50] Mukerjee S. Functional correlation of Auxin and serotonin in plants: Possible receptor surrogacy? Neurotransmitters in Plant Signaling and Communication. 2020. pp. 107-122. DOI: 10.1007/978-3-030-54478-2_6

[51] Murch SJ, Saxena PK. Role of indoleamines in regulation of morphogenesis in vitro cultures of St. John's wort (*Hypericum perforatum L*). Acta Horticulture. 2004;**2004**:425-432. DOI: 10.17660/ActaHortic.2004.629.56

[52] Ramakrishna A, Giridhar P, Ravishanker GA. Indoleamines and calcium channels influence morphogenesis in vitro culture of *Mimosa pudica* L. Plant Signaling Behaviour. 2009;**4**:1136-1141. DOI: 10.4161/psb.4.12.10101

[53] Gloria MA, Adeo RC. Effect of gamma radiation on the ripening and levels of bioactive amines in bananas cv, Prata. Radiation Physics and Chemistry. 2013;**87**:97-203. DOI: 10.1016/j.radphyschem.2013.02.032

[54] Berger M, Gray HA, Roth BL. The expanded biology of serotonin. Annual Review of Medicine. 2009;**60**:355-366. DOI: 10.1146/annurev.med.60.042307.110802

[55] Janet Best H, Nijhout F, Reed M. Serotonin synthesis, release and reuptake in terminals: A mathematical model. Theoretical Biology & Medical Modelling. 2010;**7**:34. DOI: 10.1186/1742-4682-7-34

[56] Breyon Y, Lee HY, Hwang OJ, Lee HJ, LeeK BK. Coordinated regulation of melatonin synthesis and degradation genes in rice leaves in response to cadmium treatment. Journal of Pineal Research. 2015;**58**:470-478. DOI: 10.1111/jpi.12232

[57] Gatineau F, Fouche JG, Kevers C, Hausman JF, Gaspar T. Quantitative variations of indolyl compounds

including IA, IAA-aspartate and serotonin in walnut microcutting during root induction. Biology Plant. 1997;**39**:131-137

[58] Wan J, Zhang P, Wang R, Sun L, Qiong J, Jin X. Comparative physiological responses and transcriptome analysis reveal the roles of melatonin and serotonin in regulating growth and metabolism in *Arabidopsis*. BMC Plant Biology. 2018;**18**:362. DOI: 10.1186/s12870-018-1548-2

[59] Rayne S. Concentrations and profiles of melatonin and serotonin in fruits and vegetables during fruit ripening - A mini review. Nature Proceedings. 2010. DOI: 10.1038/npre.2010.4722.1

[60] Lei Q, Wang L, Tan D-X, Zhao Y, Zheng X-D, Chen H, et al. Identification of genes for melatonin synthetic enzymes in 'Red Fuji' apple (Malus domestica Borkh. Cv. Red) and their expression and melatonin production during fruit development. Journal of Pineal Research. 2013;**55**:443-451. DOI: 10.1111/jpi.12096

[61] Zhao Y, Tan D-X, Lei Q, Chen H, Wang L, Li Q-T, et al. Melatonin and its potential biological functions in the fruits of sweet cherry. Journal of Pineal Research. 2013;**55**:79-66. DOI: 10.1111/jpi.12044

[62] Murch SJ, Hall BA, LeCH SPK. Changes in the levels of indoleamine phytochemicals during veraison and ripening of wine grapes. Journal of Pineal Research. 2010;**49**:95-100. DOI: 10.1111/j.1600-079X.2010.00774.x

[63] Roshchina VV, Melnikova EV. Allelopathy and plant reproductive cells; participation of acetylcholine and histamine in signaling in the interaction of pollen and pistil. Allelopathy Journal. 1998;**7**:171-182

[64] Fellows LE, Bell EA. 5-hydroxy-L-tryptophan, 5-hydroxytryptamine and L-Tryptophan-5-hydroxylase in *Griffonia simplicifolia*. Phytochemistry. 2022;**9**:2389-2396. DOI: 10.1016/S0031-9422(00)85745-3

[65] Hano S, Shibuya T, Imoto N, Ito A, Imanishi S, Aso H, et al. Serotonin content in fresh and processed tomatoes and its accumulation during fruit development. Scientia Horticulture. 2017;**214**:107-113. DOI: 10.1016/j.scienta.2016.11.009

[66] Murch SJ, Saxena PK. Mammalian neurohormones: Potential significance in reproductive physiology of St. John's wort (*Hypericum perforatum* L)? Naturwissenschaften. 2002;**89**:555-560. DOI: 10.1007/s00114-002-0376-1

[67] Gong D, He F, Liu J, Cheng Z, Wang Y, Tian S, et al. Understanding of hormonal regulation in rice seed germination. Life (Basel). 2022;**12**:1021. DOI: 10.3390/life12071021

[68] Azouzi S, Santuz H, Morandat S, et al. Antioxidant and membrane binding properties of serotonin protect lipids from oxidation. Biophysical Journal. 2017;**112**(9):1863-1873. DOI: 10.1016/j.bpj.2017.03.037

[69] Mukerjee S, David A, Yadev S, Baluska F, Bhatla SC. Salt stress induced seedling growth inhibition coincides with differential distribution of serotonin and melatonin in sunflower seedling roots and cotyledons. Physiologia Plantarum. 2014;**152**:714-728. DOI: 10.1111/ppl.12218

[70] Shi H, Ye T, Chan Z. Comparative proteomic responses of two bermudagrass (*Cynodon dactylon* (L). Pers.) varieties contrasting in drought stress resistance. Plant Physiology Biochemical. 2014;**82**:218-228. DOI: 10.1016/j.plaphy.2014.06.006

[71] Arnao MB, Hernandez-Ruiz J. The physiological function of melatonin in plants. Plant Signaling & Behavior. 2006;**1**:89-95. DOI: 10.4161/psb.1.3.2640

[72] Baluska F, Mancuso S. Root apex transition zone as oscillatory zone. Frontiers in Plant Science. 2013;**4**:354. DOI: 10.3389/fpls.2013.00354

[73] Ishihara A, Hashimo Y, Tanaka C, Dubouzet JG, Nakao T, Matsuda F, et al. The tryptophan pathway is involved in the defense responses of rice against pathogenic infection via serotonin production. The Plant Journal. 2008;**54**:481-495. DOI: 10.1111/j.1365-313X.2008.03441.x

[74] Dharmawardhana P, Ren L, Amarsinghe V, Monaco M, Thomson J, Ravenscroft D, et al. A genome scale metabolic network for ice and accompanying analysis of tryptophan, auxin and serotonin biosynthesis regulation under biotic stress. Rice. 2013;**6**(1):15. DOI: 10.1186/1939-8433-6-15

[75] Liu Y, Ding X, Lv Y, Cheng Y, Li C, Yan L, et al. Exogenous serotonin improves salt tolerance in rapeseed (*Brassica napus* L) seedlings. Agronomy. 2022;**11**:2. DOI: 10.33390/agronomy11020400

[76] Vanitha PA, Vijayaraghavareddy P, Uttarkar A, et al. Nevel small molecules targeting bZIP TF improve stomatal conductance and photosynthesis and photosynthesis under mild drought stress by regulating ABA. FEBS Journal. 2022;**289**:6058. DOI: 10.1111/febs.16461

[77] Akcay UC, Okudan N. Exogenous serotonin improves drought and salt tolerance in tomato seedlings. Research Square. 2023;**101**:239-249. DOI: 10.21203/rs-2714889/v1

[78] Kumar G, Saad KR, Arya M, Puthusseri B, Mahadevappa P, Shetty NP, et al. The synergistic role of serotonin and melatonin during temperature stress in promoting cell devision, ethylene and isoflavones biosynthesis in Glycin max. Current Plant Biology. 2021;**26**:100206. DOI: 10.1016/j.cpb.2021.100206

[79] Arland LAE, Turi CE, Saxena PK. Chapter 2 - Serotonin in plants: Origin, functions and implications. Serotonin. 2019:23-46. DOI: 10.1016/B978-0-12-800050-2.00002-4

[80] Akcay UC, Okudan N. Exogenous serotonin improves drought and salt tolerance in tomato seedlings. Research Square. 2023. DOI: 10.21203/rs.3.rs-2714889/v1

[81] Huang YM, Kao CH. Calcium in the regulation of corn leaf senescence by light. Botanical Bulletin of Academia Sinica. 1992;**33**:161-165

[82] Lazar D, Murch SJ, Beilby MJ, Khazaaly AIS. Exogenous melatonin affects photosynthesis in characeae *Chara australis*. Plant Signaling & Behavior. 2013;**8**:e23279. DOI: 10.4161/psb.23279

[83] Roshchina VV. Regulation of chloroplast reaction by secondary metabolite acetylcholine and biogenic amines. Acta Botanica Croatica. 1990;**49**:29-35

[84] Conn PJ, Sanders-Bush E. Biochemical characterization of serotonin stimulated phophoinositide turnover. Life Science. 1986;**17**(3):663-669. DOI: 10.1.16/0024-3205(86)90060-3

[85] Claustre Y, Rouquier L, Scatton B. Pharmacological characterization of serotonin-stimulated phosphoinositide turnover in brain region of the immature rat. The Journal of Pharmacology

and Experimental Therapeutics. 1988;**244**(3):1051-1056

[86] Das R, Sopory SK. Evidence of regulation of calcium uptake by phytochrome in maize protoplasts. Biochemical and Biophysical Research Communication. 1985;**128**:1455-1460

[87] Mannino G, Pernici C, Serio G, Gentile C, Bertea CM. Melatonin and phytomelatonin: Chemistry, biosynthesis, metabolism and bioactivity in plants and animals—An overview. International Journal of Molecular Sciences. 2021;**22**(18):9996. DOI: 10.3390/ijms22189996

[88] Ramakarishna A, Ravishanker GA. Influence of abiotic stress signals on secondary metabolites in plants. Plant Signaling & Behavior. 2011;**1**(6):1720-1731. DOI: 10.4161/psb.6.11.17613 doi: 10.4161/psb.6.11.17613

[89] Fankhauser C. Photomorphogenesis in Plants: Genetics of Brenner's Encyclopedia of Genetics. 2nd ed. San Diego: Academic Press; 2013. pp. 309-312. DOI: 10.1016/B978-0-12-374984-0.01156-6

[90] Bajwa VS, Shukla MR, Sherif SM, Murch SJ, Saxena PK. Identification and characterization of serotonin as an anti-browning compound of apple and pear. Postharvest Biology and Technology. 2015;**110**:183-189. DOI: 10.1016/j.postharvbio.2015.08.018

[91] Wolfgang G. Function of serotonin in seeds of walnut. Phytochemistry. 1982;**21**(4):819-822. DOI: 10.1016/0031-9422982080071-X

Chapter 7

Serotonin in the Nervous System: Few Neurons Regulating Many Functions

Citlali Trueta and Montserrat G. Cercós

Abstract

Serotonin is synthesized from tryptophan in small groups of neurons within the central nervous system. These neurons, however, branch profusely and innervate all the nervous system, where, by releasing serotonin in different manners, they regulate a myriad of functions, including many behaviors. This chapter reviews the main functions of serotonin in the nervous system of invertebrates and vertebrates, showing that many of these have been conserved throughout evolution. It also summarizes the current knowledge about the mechanisms that control and regulate serotonin secretion from different compartments of the same neurons, evidencing their differences, which enable small numbers of neurons to display a wide variety of functions, including the regulation of our mood states.

Keywords: serotonin, 5-HT, synapse, extrasynaptic secretion, neuromodulation, behavior, central nervous system

1. Introduction

From the mid-nineteenth century, a substance in the blood serum was known to cause contraction of smooth muscle, regulating the "tone" of blood vessels. When this substance was isolated, it was called serotonin, for its first known function. Similarly, a substance that causes smooth muscle contraction in the digestive tract, secreted by enterochromaffin cells, was called "enteramine" [1, 2]. Both substances were later demonstrated to be the same molecule: 5-hydroxytryptamine (5-HT).

Serotonin, or 5-hydroxytryptamine is a monoamine that acts as a chemical messenger, both in and out of the nervous system of invertebrates and vertebrates. It is one of the most ancient molecules regulating cellular functions, since it is found from protozoans [3]. Serotonin regulates a wide variety of physiological functions, from those already mentioned on smooth muscle contraction to very complex ones, such as attention or social behavior. Serotonin is secreted by exocytosis from the cells that synthesize it and can act as a neurotransmitter, neuromodulator, or hormone.

Most of the serotonin in mammals is found in cells outside of the brain, such as platelets, mastocytes, or enterochromaffin cells. However, serotonin cannot go through the blood-brain barrier, and thus, cerebral serotonin must be synthesized in the brain. Within the nervous system, serotonin has multiple functions, acting as a neurotransmitter at synapses or as a neuromodulator both at synapses and at extra-synaptic sites. Its functions are notably well conserved throughout the phylogenetical scale, as are the basic characteristics of serotonergic systems, which are characterized by having small numbers of neurons that, nevertheless, act at multiple levels in the nervous system, regulating many different functions.

2. Metabolism and effects of serotonin in the nervous system

2.1 Synthesis and release

Serotonin is synthesized in the cytoplasm of neurons, from the amino acid L-tryptophan, which is captured by cells through a process of facilitated transport. The main source of tryptophan are the proteins obtained from the diet. The enzyme tryptophan hydroxylase transfers an oxygen atom to the fifth position of the ring, forming 5-hydroxytriptophan. For this, the enzyme requires molecular oxygen (O_2), as well as reduced pteridine as a cofactor. This reaction is the limiting step in serotonin synthesis. Since in the brain the concentration of tryptophan normally does not saturate the enzyme, serotonin synthesis is increased by increases in the ingest of this amino acid. The level of O_2 in the tissues also regulates the synthesis of serotonin. 5-hydroxytriptophan is immediately decarboxylated by the enzyme aromatic amino acid decarboxylase, to form 5-hydroxytryptamine or serotonin. In addition to the effects of serotonin, which will be discussed below, this molecule serves as a precursor for the hormone melatonin in the pineal gland and other tissues. Serotonin synthesis can be increased in situations that require its continuous release. For example, electrical stimulation of the serotonergic neurons increases serotonin synthesis in a frequency-dependent manner. This can occur without increasing the synthesis of tryptophan hydroxylase, probably by changing the kinetic properties of the enzyme by phosphorylation through calcium-dependent mechanisms [4].

Serotonin release occurs through exocytosis, since it is a charged molecule that cannot go through the plasma membrane. In addition, it is necessary to store serotonin in vesicles to protect it from degradation (see below). The rate of serotonin release depends on the firing frequency of the neurons that synthesize it. The mechanisms that regulate serotonin release will be described further below in this chapter.

2.2 Reuptake and degradation

The activity of serotonin is terminated principally by its reuptake in serotonergic neurons and in glial cells, through a membrane transporter. Reuptake is an active process, which depends on the temperature and of extracellular sodium and chloride. Energy is required to maintain the sodium gradient between the intra and extracellular media, which is indispensable for serotonin transport. Serotonin reuptake by serotonin transporter (SERT) may produce changes in the membrane potential, since it moves positively charged sodium ions into the cytoplasm, thus slightly depolarizing the neuron [5]. Many of the pharmacological treatments for depression are selective serotonin reuptake inhibitors (SSRIs).

Serotonin is degraded by the enzyme monoamine oxidase (MAO) that converts 5-HT to 5-hydroxindoleacetaldehyde, which in turn can be oxidized by a

NAD^+-dependent aldehyde dehydrogenase to form 5-hydroxindolacetic acid or reduced by a NADH-dependent aldehyde reductase to produce 5-hydroxy tryptophol, depending on the relative concentrations of NAD^+ and NADH in the tissue. In the brain, the main metabolite of serotonin is 5-hydroxindolacetic acid.

2.3 Cellular effects of serotonin on the nervous system

Serotonin acts on seven groups of receptors, named 5-HT_1–5-HT_7, distinguished by their pharmacology and the intracellular pathways and responses they activate in the target cells. Each receptor type is differentially distributed in different areas of the nervous system and other tissues. All serotonin receptors, except the 5-HT_3 (which is an ionic channel), belong to the family of G-protein-coupled receptors, similar to rhodopsin, and activate intracellular signaling cascades. The effects that serotonin exerts on target neurons depend on the type of receptor activated. The G_i/G_o-coupled 5-HT_1 receptors generally mediate inhibitory effects on neuronal firing through an opening of inwardly rectifying potassium channels or a closing of voltage-gated calcium channels. The $G_{q/11}$-coupled 5-HT_2 family of receptors generally mediates slow excitatory effects through a decrease in the membrane permeability for potassium or an increase in the permeability for cations. The 5-HT_3 receptors, which are ligand-gated cation channels with structural homology to nicotinic receptors for acetylcholine, mediate fast excitatory effects of 5-HT. In invertebrates there seem to be ionotropic 5-HT receptors [6–8], which are channels for chloride (a negatively charged ion), unlike the vertebrate 5-HT_3 receptor, which allows the flux of positively charged ions.

The characteristics, signaling pathways, and distribution of each of the receptors for serotonin have been extensively reviewed elsewhere [9–13] and will not be further discussed here.

In this chapter, we provide an overview of the morphology and function of serotonergic neural systems, and show how, by secreting this substance in several different modes, a few neurons regulate a wide variety of functions in the nervous system.

3. The diverse functions of serotonin in the nervous system

The functions of serotonin have been well conserved along evolution, from invertebrates to humans. In addition to developmental, cardiovascular, gastrointestinal, and endocrine functions, serotonin has multiple functions in the nervous system, including the regulation of sensory perception and motor patterns (such as those that are activated during ingestion, locomotion, or respiration) [14–16], circadian rhythms, including sleep-wake cycles [17, 18], appetite, feeding, sexual behavior, mood, attention, cognition, and memory [19]. Importantly, serotonin regulates social behavior in invertebrates and vertebrates, and produces changes that span from aggression, associated with social dominance [20], to submission and depression. In humans, the alteration in serotonin metabolism is related to neuropsychiatric disorders, such as depression [21], schizophrenia [22, 23], and obsessive-compulsive [23, 24] or feeding disorders [25].

3.1 Serotonergic functions in invertebrates

The relatively small and simple nervous systems of invertebrates, with small numbers of neurons of big size and stereotyped localization, have been powerful model systems to study neural circuits in detail, from intracellular pathways to behavior. Serotonin regulates a wide variety of physiological functions in invertebrates, including sensory processing, locomotion, feeding, and social behavior. The accessibility of invertebrate nervous systems to perform intracellular recordings of identified neurons has enabled us to unveil complete circuits regulating these behaviors and the multiple ways in which serotonin modulates them.

3.1.1 Regulation of swimming, feeding, and learning in leeches

Serotonin released from Retzius neurons has a neuromodulatory effect on the block of conduction of action potentials that occurs at the branching points of the axons of mechanosensory neurons [26] that respond to touch (T), pressure (P), or noxious (N) stimuli in the skin of the leech [27]. In this way, serotonin modulates sensory perception of these animals.

Serotonin also regulates a wide variety of physiological functions in the leech, many of which are related to feeding behavior. Hungry leeches are usually located in shallow waters, and respond to water movements, possibly indicating the presence of prey, by swimming toward the source of the waves. Serotonergic neurons play a fundamental role in swim initiation. Swimming in these organisms is coordinated by a series of interneurons that, when activated, display an oscillatory activity that stimulates and inhibits, in a rhythmic and alternative manner, the motor neurons that excite and inhibit longitudinal ventral and dorsal muscles. Activation of this central pattern occurs progressively through the ganglion chain, giving rise to undulations of the body from head to tail [28]. The rhythmic activity of these neurons can be recorded in the segmental ganglia or the connective nerve and is maintained in isolated ganglion chains, which has facilitated the study of these circuits. Mechanosensory neurons excite several of the serotonergic neurons, including Retzius neurons, cells 21 and 61 (see Section 4.1 on the Structure of the serotonergic system in invertebrates below), through a polysynaptic pathway (i.e., with one or more interneurons connected in series to each other in between) [29]. Stimulation of cells 21 and/or 61 triggers the initiation of swimming episodes in ganglion chains [29], since these cells, in turn, excite interneurons that generate the central swimming pattern, receiving also feedback from them [30]. Stimulation of Retzius neurons or application of serotonin in the bath also produces swimming episodes [31], because serotonin modulates the activity of some of the oscillating interneurons that may initiate the rhythmic activity of the swimming circuit. For example, through modulating several sodium channels, serotonin changes the excitability (i.e., the likeliness to fire action potentials) of one of the interneurons that initiate the swimming cycle (called cell 204), decreasing the threshold for this neuron to trigger the activity of the circuit [32, 33]. Serotonin also modulates the activity of motor neurons that produce swimming, promoting their rhythmic alternated activation [34, 35]. Thus, serotonin seems to be the determinant factor for the activation of swimming in the leech. In fact, the probability of swimming in a leech is correlated with the concentration of serotonin in its blood [31] and in the CNS [36].

Locomotion is also regulated by serotonin in other animals. In larvae of the fruit fly *Drosophila*, increasing the level of 5-HT is able to decrease body wall contractions used for locomotion. In the larval stages, 5-HT is involved in turning behavior [37], whereas in adult flies, acute activation of 5-HT neurons disrupts normal locomotor activity [38].

Once a leech finds possible prey, it starts an exploration phase. During swimming and this exploratory phase, the activity of serotonergic neurons increases [39]. When the lips feel a warm surface, some of the serotonergic neurons are activated, firing in bursts at high frequencies, and serotonin activates ingestion, directly stimulating salivary glands to secrete saliva, the claws to bite, and the pharynx to generate peristaltic movements, which continue throughout the ingestion process. During ingestion, serotonin relaxes the body wall muscles by producing a hyperpolarization through the activation of chloride currents during the compound action potential in these muscles [40], and by inhibiting the production of excitatory synaptic potentials in the muscles through its action on central neurons [41]. In addition, Retzius neurons directly stimulate mucus secretion at the mucous glands in the skin [42], which is also a characteristic of the ingestion phase in the leech.

The distention of the body wall after feeding hyperpolarizes the serotonergic neurons and terminates ingestion, producing a satiety state. Satiated leeches have a lower amount of serotonin in the CNS and in blood than hungry leeches [43] and stay in deeper waters at the bottom of the ponds, swimming much less. Administration or elimination of serotonin cause alterations in feeding behavior, such as hyperphagia or anorexia, respectively [44].

Beyond feeding and swimming, serotonin modulates learning processes in the leech, such as facilitation, sensitization, and dishabituation, in the shortening reflex that occurs when touching the head of the animal [45, 46]. This modulation is exerted upon the excitability of a series of neurons that are essential for learning in these organisms, called S cells [47]. The activity of the S cell network all along the nervous system is modulated by serotonin [48].

3.1.2 Regulation of swimming and sensory learning in mollusks

Serotonin regulates swimming also in mollusks. In *Clione limacine,* the serotonergic system modulates the expression of different behaviors by producing different swim velocities [49]. This is achieved by independently regulating the contractility of the wings, the cycling frequency of the central pattern generator, recruiting motor neurons for swimming, and by stimulating the heart-exciting neuron, while inhibiting a circuit that competes with the swimming circuit.

In *Aplysia californica,* serotonin has been shown to regulate learning. This mollusk has a reflex that withdraws the gill and the siphon in response to a tactile stimulus in the tail. The strength and duration of this reflex can be enhanced by a noxious electrical stimulation applied to the tail, the head, or the body wall. This sensitization phenomenon, which can last for hours, is caused by (1) an increase in the excitability of sensory neurons and (2) a facilitation of transmitter release from these neurons to the motor neurons that produce the contraction of the mantle muscles [50]. Both of these mechanisms are activated by serotonin. Noxious stimulation of the tail activates serotonin release in the CNS, which can be detected by electrochemical methods, and lasts 30–40 seconds, peaking in the neuropil (i.e., the net of neuronal processes) surrounding the synapses between sensory and motor neurons in the tail [51]. Serotonin levels also increase in the hemolymph [52]. Serotonin acts on sensory neurons to activate the protein kinase A (PKA) and protein kinase C (PKC) intracellular pathways, which phosphorylate S-type potassium channels, to reduce the time they remain open [53]. In this way, the membrane excitability increases and the action potential is broadened [54, 55], allowing more calcium influx to the nerve terminals and therefore increasing transmitter release. PKA and PKC also increase transmitter release by enhancing the mobilization of synaptic vesicles to the active zone [56–61], thus increasing the transmitter available for release. By increasing transmitter release, a stronger contraction is produced in the muscle that withdraws the siphon. The modulation of defensive reflex circuits by 5-HT also involves a variety of different cell types that respond differently to 5-HT. Serotonin blocks the synaptic enhancement in a polysynaptic pathway that inhibits the siphon-withdrawal reflex, thereby reducing potential for inhibition within the circuit. Serotonin also acts directly on motor neurons facilitating responses to glutamate mediated by AMPA-type receptors, thus possibly enhancing the response of motor neurons to sensory neuron activation (Reviewed by [62]).

3.1.3 Regulation of aggression associated with social behavior in arthropods

The modulation that serotonin exerts on behaviors associated with social hierarchy has been studied in detail in decapod crustaceans, such as lobsters and crayfish. When two individuals meet at the same space, at the beginning both of them show aggressive behavior, consisting of standing high on the tips of their walking legs, while bending them, and showing their claws up and forward. Both animals have aggressive encounters, which will determine which one will be dominant. The dominant individual then keeps a high posture with the limbs flexed, and walks throughout the space, while the other one tends to extend its limbs, to place the body low, near the bottom, and remains in the corners (when confined in a tank), or leaves the area (in the wild), avoiding the dominant individual. Serotonin, together with octopamine, is responsible for these opposite postures and behaviors. Serotonin injection to lobsters produces the dominant posture with flexed limbs and abdomen, while octopamine produces the submission behavior with the extension of the limbs [20, 63]. Serotonin activates a motor program for the flexion of the limbs, by increasing the firing frequency of the motor neurons that excite the flexor muscles and those that inhibit the extensor muscles while decreasing the firing frequency of the motor neurons that inhibit the flexor muscles and of those that excite the extensor muscles. This occurs by modifying the frequency of the synaptic inputs to the motor neurons, and not by acting directly upon them [64]. Octopamine produces the opposite effects [64]. Serotonin also acts directly on the muscles, by producing prolonged contractions, and on the neuromuscular junctions (i.e., the synapses between motor neurons and muscles), by increasing neurotransmitter release and therefore the strength of muscle contractions [65–67]. These effects are produced by serotonin released from two neurosecretory organs to the hemolymph [68]: (see Section 4.1 on the Structure of the serotonergic system in invertebrates) where it circulates as a neurohormone and regulates different targets, including the exoskeleton muscles and the heart.

The role of serotonin in aggressive behavior associated with social dominance has also been studied in the fruit fly *Drosophila melanogaster.* The multiple genetic manipulation methods developed in this organism have been powerful tools for studying a variety of functions. Recent tools to manipulate the biosynthesis and release of serotonin, the activity of serotonergic neurons, or the expression of 5-HT receptors, have revealed the role of serotonin in a variety of behaviors in these animals. As in lobsters, *Drosophila* males establish hierarchical relationships based on the results of aggressive encounters. These fights have several stages in which the animals show a number of characterized behaviors, including tapping on the opponent's leg, lunging, holding, boxing, or tussling. The result of these encounters defines a dominant individual, who takes a food source, while the defeated individual retreats from the food surface [69, 70]. Increasing the level of serotonin in the fly's nervous system, either by increasing 5-hydroxytryptophan in the diet or by constitutive over-expression of tryptophan hydroxylase in serotonergic neurons, increases aggressive behavior [71]. Similarly, when serotonergic neurons are selectively activated by expressing cationic channels that can be activated by temperature in these cells, aggression is expressed faster, fights become intensified and continue even after social dominance has been established. By contrast, the reduction of serotonergic neurotransmission by the selective expression of a temperature-sensitive form of dynamin, a protein involved in endocytosis, in serotonergic neurons, results in a reduction in male mid-intensity aggression and the development of fewer dominance relationships [72].

3.1.4 Regulation of circadian rhythms in invertebrates

The relationship between circadian rhythm and sleep-wake has been also studied in invertebrates. In *Drosophila*, modifying the serotonin system can break the link between sleep and the circadian rhythm. In fact, an increase in serotonin reinforces sleep, presumably *via* the 5-HT_{1A} receptor, while a decrease in serotonin by the deletion of tryptophan hydroxylase suppresses sleep at night [73]. In crustaceans, the regulation of circadian rhythms is operated in both the central brain and the eye-stalk X-organ-sinus gland system, where serotonin and melatonin have been largely found [74]. In crayfish, serotonin concentration in tissue displays circadian fluctuations, linking the serotonin system with the pacemaker system involved in circadian rhythms [75].

3.2 Serotonergic functions in vertebrates

Serotonin regulates a wide variety of functions and behaviors also in vertebrates, many of which are conserved along the phylogenetic scale. However, in contrast to invertebrates, where the pathways and mechanisms through which serotonin acts are very well known, the precise pathways and mechanisms of serotonergic effects in vertebrates remain elusive, due to the complexity of the vertebrate nervous system.

3.2.1 Regulation of locomotion

As in invertebrates, central 5-HT is also a powerful neuromodulator of locomotor activities in vertebrates, including lamprey [76–79], zebrafish [80–83], Xenopus [84, 85], and rodents [86–88]. Serotonergic neurons terminate on specific target neurons with different types of serotonergic receptors in the spinal cord. Serotonergic neurons can initiate locomotor activity through actions on motoneurons and neurons of the locomotor central pattern generator (CPG) [89, 90]. In fact, the stimulation of serotonergic neurons projecting to the spinal cord, or application of 5-HT receptor agonists on *ex vivo* spinal cord from newborn rats, initiates and sustains episodes of so-called fictive locomotion [91, 92]. In rats and mice, serotonergic agonists promote the recovery of locomotor movements after spinal cord injury [89].

3.2.2 Regulation of feeding

As in invertebrates, serotonin regulates feeding in vertebrates. However, while in invertebrates serotonin promotes appetitive states, in vertebrates serotonin suppresses appetite [93]. In addition to the peripheral effects on metabolism through stimulating insulin and inhibiting glucagon secretion (reviewed in a different chapter), serotonin has important actions in the regulation of food intake at the central nervous system (CNS) level. Food ingestion is regulated in the hypothalamus, by two types of neurons in the arcuate nucleus: one that releases agouti-related peptide (AgRP) and neuropeptide Y (NPY) and one that releases alpha-melanocyte stimulating hormone (α-MSH), also called POMC neurons, because they express pro-opiomelanocortin. Both of these innervate neurons in the paraventricular nucleus, which express melanocortin MC4 receptors and, when activated, inhibit food ingestion. α-MSH is an agonist of melanocortin receptors, while AgRP/NPY is an antagonist of these receptors. In addition, NYP neurons release GABA onto POMC neurons, inhibiting them. Thus, activation of POMC neurons results in a decrease in food intake, while activation of NPY/AgRP neurons results in an increase in food intake. Serotonin produces decreases food intake, by depolarizing POMC neurons and by inhibiting NYP neurons. Depolarization of POMC neurons occurs through the activation of 5-HT_{2C} receptors, by inhibiting both GIRK and M potassium channels, which participate in the maintenance of the resting potential [94], and activating TRPC channels, thus decreasing their input resistance [95]. Hyperpolarization of NPY/AgRP neurons by activation of 5-HT_{1B} receptors decreases the release of AgRP/NPY onto paraventricular nucleus neurons and also decreases inhibitory synaptic input onto POMC neurons [96]. In consequence, inhibition of serotonin synthesis and release increases food intake and body weight in rats [97, 98]. Moreover, mutant mice that do not express serotonergic 5-HT_{2C} receptors show increased food intake and increased body weight and adipose tissue than wild-type mice [99, 100]. On the other hand, the treatment with fluoxetine [101], a selective serotonin reuptake inhibitor (SSRI), or with d-Fenfluramine [96], which inhibits serotonin reuptake and stimulates its secretion, decreases food intake in wild-type rodents, but not in mice lacking 5-HT_{2C} receptors [102]. Thus, serotonin in vertebrates decreases appetite and feeding. This is the main difference between the functions of serotonin in vertebrates and those in invertebrates, where serotonin stimulates feeding. The hypophagic effects of serotonin in humans have been related to feeding disorders, such as anorexia and bulimia [25].

3.2.3 Serotonergic regulation of circadian rhythms and sleep

In mammals, 5-HT is implicated in sleep-wake states. Studies using electrophysiological, neurochemical, genetic, and neuropharmacological approaches have shown that serotonin promotes wakefulness and inhibits rapid eye moment (REM) sleep. Indeed, mutant mice that do not express 5-HT_{1A} receptors exhibit greater amounts of REM sleep than their wild-type counterparts. Recordings from serotoninergic neurons in unanesthetized animals have shown that activity is highest during periods of waking arousal, reduced in quiet waking, reduced further in slow-wave sleep, and absent during REM sleep [103]. However, under certain circumstances, this neurotransmitter contributes to the increase in sleep propensity [104]. Thus, serotonergic activity may be accompanied by waking or sleep depending on the brain area and receptor type involved in the response and also on the concomitant agonism/antagonism of other neurotransmitter systems [105].

Serotonin also participates in the control of the circadian rhythms. This participation is supported by the fact that there is a significant projection of serotonergic neurons form the raphe nuclei to the suprachiasmatic nucleus, which is considered the master clock regulating circadian rhythms [106]. Also, one of the metabolites of serotonin is melatonin, a molecule that is known to regulate the sleep-wake cycle. In doves, serotonin levels in serum seem to be positively correlated with the circadian activity rhythm [107]. In mammals, it is well established that serotonin modulates the sensitivity of the circadian rhythm to light through the modulation of a presynaptic 5-HT_{1B} receptor on the retino-hypothalamic tract; the activation of this receptor attenuates photic input to the central nervous system, thereby reducing the phase response to light [106].

3.2.4 Regulation of aggression associated with social behavior in mammals

Aggressive behavior and the establishment of social dominance are also modulated by serotonin in vertebrates. In mammals, including humans, serotonin inhibits aggressive behavior by blocking the secretion of vasopressin and other transmitters [108, 109]. Serotonergic and vasopressinergic innervation of the hypothalamus can be altered in hamsters that are exposed to aggression during adolescence, resulting in altered aggressive behavior in the adult [108]. The activity of serotonergic neurons in the rat CNS determines the transition between normal and escalated types of aggression [110]. Mutant mice that do not express 5-HT_{1B} receptors show increased aggression, as well as auto administration of cocaine and ethanol [111].

Several lines of evidence point to low levels of serotonin in the cerebrospinal fluid and in specific areas of the brain in individuals with increased aggressive behavior, both in rodents [112–115] and in humans [116, 117]. In contrast, higher levels of serotonin are related to adaptive social behaviors [118, 119]. In primates, individuals with a low serotonergic activity in the CNS [120], or treated with 5-HT antagonists [121], express impulsive, aggressive, and social isolation behaviors, and invariably become subordinates, while individuals treated with 5-HT agonists or 5-HT reuptake inhibitors show more affiliation behaviors and less aggressivity, and become dominants [121].

3.2.5 Participation of serotonin in mood, mental health, and other behaviors

Studying the neurobiological bases of mood disorders faces a number of difficulties, because the brain is not an approachable tissue in humans, and research on this topic has relied upon post-mortem studies, or neuroimaging of living subjects, both of which have great limitations. There are a number of animal models too, with concomitant challenges for their interpretation in terms of human mental health. However, several lines of evidence support that serotonin plays an important role in mood states, and is implicated in mental health disorders, such as depression, anxiety, or schizophrenia, among others [122].

Patients with depression and epilepsy have a deficit in serotonergic transmission [123]. Human clinical studies have employed a range of serotonin indexes, including the cerebrospinal fluid level of serotonin metabolites and the prolactin response to serotonin agonists. A positive correlation was found between low levels of serotonin metabolites in the cerebrospinal fluid and serious or high-intent suicidal acts [124, 125].

Serotonin transporter (SERT) is the main target of SSRI antidepressants. However, although the therapeutic effects induced by SSRI are initially triggered by blocking SERT, they rely on consequences of chronic exposure [126], including desensitization of somatodendritic 5-HT autoreceptors. One of the events that seem to mediate antidepressant effects is hippocampal neurogenesis, which is negatively regulated by stress and positively regulated by antidepressant treatment [127].

Exposure to chronic unpredictable stress has been found to induce depressive-like symptoms or behaviors, including passive behavioral coping and anhedonia in animal models, along with some other affective, cognitive behavioral symptoms that are also present in humans. In models of chronic unpredictable stress, it has been shown that serotonergic activity and neurotransmission, as well as autoreceptor sensitivity are altered [127, 128]. Along the same line, the pathway from the medial raphe nucleus to the hippocampus attenuates stress through facilitating transmission in that area [129].

It has been proposed that serotonergic pathways in mammals regulate anxiety. Activation of the ascending pathway of the dorsal raphe nucleus (DRN) facilitates defensive learned behaviors. On the other hand, activation of the pathway from the DRN to the periventricular area inhibits innate "fight or flight" reactions. It is thought that the disfunction of these pathways is related to generalized anxiety disorder and panic [130]. The pathway from the medial raphe nucleus to the hippocampus, on the other hand, attenuates stress through facilitating transmission in that area [129]. Mice lacking 5-HT_{1A} receptors show increased fear in several behavioral tests, suggesting that this receptor modulates some neural circuits related to fear [131]. In fact, 5-HT_{1A} agonists are used, in addition to benzodiazepines (which increase GABAergic activity), to treat generalized anxiety disorders [132]. Inhibition of the serotonin transporter can also reduce anxiety symptoms in obsessive-compulsive disorder [24].

Sexual behavior is also modulated by serotonin, possibly through the regulation of dopamine secretion in the hypothalamus. Serotonin inhibits some aspects of sexual behavior in both male and female rats; it increases the latency to copulation and ejaculation [133], as well as the refractory period between ejaculation and the next copulation in males, and decreases female receptivity (reviewed by Weiger WA [109]).

In vertebrates, serotonin also modulates pain. Descending pathways from the raphe nuclei to the spinal cord are activated upon peripheral injury and produce an inhibitory effect on pain perception [134]. It is interesting to note that some patients with pain disorders also present mood symptoms.

Serotonergic neurotransmission is altered in schizophrenia [135], and the treatment of this disease includes drugs that inhibit 5-HT_2 as well as dopaminergic receptors. Finally, some serotonergic 5-HT_{1D} agonists are successfully used in the treatment of migraine [136].

4. The serotonergic system: few neurons releasing serotonin in multiple ways

A striking characteristic of serotonergic systems is that they, in general, have small numbers of neurons in relation to the total neurons in the nervous system. This is conserved throughout the phylogenetic scale, from invertebrates to mammals. In the leech, for example, there are seven serotonergic neurons, out of 400 total neurons in each segmental ganglion. In rodents, there are around 9000 serotonergic out of 10^{12} total neurons. In the human brain, the number of serotonergic neurons represents only one out of every million neurons [137]. Yet, serotonergic projections in vertebrates and invertebrates branch profusely and have complex innervations to virtually all areas of the central nervous system, and serotonin regulates a wide variety of functions, from the modulation of inputs at sensory systems to the performance of complex behaviors.

4.1 Structure of the serotonergic system in invertebrates

The large and identifiable neurons of invertebrates, in which nervous systems have small numbers of neurons compared to those in vertebrates, provide a unique accessibility to study neural circuits at the cellular level, allowing to trace the actions of serotonin throughout circuits and to study the functions of serotonin from the cellular level to behavior.

In invertebrates, including annelids, mollusks, and crustacea, where the nervous system is formed by a ventral chain of ganglia, the serotonergic system is comprised of a few neurons in each neural ganglion. In leeches, there are seven serotonergic neurons in each segmental ganglion [138]: the pair of Retzius neurons, which are the biggest neurons in the ganglion and the ones that contain most of the serotonin in this system, two pairs of lateral neurons (one in the dorsal side of the ganglion, also called cells 21, and one on the ventral side, also called cells 61), and an unpaired neuron in the medial posterior package. The three first ganglia have an additional pair, near the pair of Retzius neurons. Because of their large size, Retzius neurons have been used to study the detailed characteristics of serotonin secretion. Retzius neurons send axons toward the periphery, through lateral roots in each ganglion, while the rest of serotonergic neurons have their dendritic arbor restricted within the central nervous system (CNS). All these neurons are coupled among them by chemical excitatory synapses as well as by electrical synapses that allow the flux of current in both directions. In addition, they receive common (apparently cholinergic) synaptic inputs. Thus, all serotonergic neurons display somewhat synchronous electrical activity in bursts and form a "compartmental serotonergic system," embedded in motor networks controlling feeding, escape swim/turn, and locomotor functions [49, 93]. Serotonergic neurons in these animals establish some synaptic contacts with other neurons in these networks and receive reciprocal innervation from the networks they innervate as well as external afferents. In addition, as will be explained below, these neurons contain large groups of serotonin-containing dense-core vesicles in the soma and at extrasynaptic sites of the axons.

In the lobster, there is at least one pair of serotonergic neurons in each ganglion. Two of these pairs, located in the fifth thoracic ganglion (T5) and the first abdominal ganglion (A1) are responsible for the secretion of serotonin in the neuropile and to the periphery at neurosecretory organs. These neurons are stimulated by the command interneurons that activate the flexor motor pattern of the legs (see Section 3.1 on Serotonergic functions in invertebrates), and serotonin then acts as an amplifier of the activation signal from the motor pattern, producing the effects already mentioned on motor neurons, muscles, and neuromuscular junctions [20]. On the other hand, when the motor pattern for the extension of the legs is activated, these serotonergic neurons are inhibited.

In the mollusk *Clione limacine,* there are 27 pairs of serotonergic neurons, 75% of which have been identified. The serotonergic system in this organism is compartmentalized, so that each subsystem can act independently or in synchrony with the others to produce variability in the speed of locomotion, thus modulating the expression of different behaviors. A cluster of podal 5-TH neurons increases the contractility of the wings without affecting their beating frequency or the activity of motoneurons. Two clusters of cerebral 5-HT neurons produce responses that increase the cycling frequency of the central pattern generator, recruit motoneurons for swimming, activate the podal 5-HT neurons, and excite the heart-exciting neuron. On the other hand, a pair of cerebral 5-HT neurons exerts a weak excitatory input to the swimming circuit and strongly inhibits neurons in another circuit, which competes with the swimming circuit [49].

4.2 Structure of the serotonergic system in vertebrates

In 1964, Dahlstrom and Fuxe found, using the Falck-Hillarp technique of histofluorescence, that most of the somata of serotonergic neurons are grouped in nine clusters localized at the midline of the brain stem, which previously had been designated as the raphe nuclei, based on cell body structural characteristics and organization, and they named B_1 through B_9 [139], although some serotonergic neurons are out of these nuclei and not all the neurons in the raphe nuclei are serotonergic. Therefore, unlike invertebrates, where serotonergic neurons are distributed throughout the nervous system in the different ganglia, vertebrates concentrate the serotonin neuron somata in a few nuclei in the brainstem. However, despite the restricted localization of the somata to this area, brainstem serotonin neurons send ascending projections that branch profusely and terminate in a defined and organized manner in cortical, limbic, midbrain, and hindbrain regions, as well as descending projections to the spinal cord, and thus the axons of serotonergic neurons innervate virtually all areas of the CNS (**Figure 1**). Groups B1 to B5, which are small and are in the most caudal raphe nuclei, send projections within the brainstem and toward the spinal cord, where they modulate the activity of motoneurons as well as synaptic transmission in the pain perception pathway. Group B7 (the largest group of serotonergic cells), which together with group B6 constitutes the dorsal raphe nucleus, B8, which corresponds to the medial or central superior nucleus, and B9, which is not considered one of the raphe nuclei, innervates all the forebrain, including the cortex. The two main ascending serotonergic pathways emerging from the midbrain raphe nuclei to the forebrain, named the dorsal periventricular path and the ventral tegmental radiations, converge in the caudal hypothalamus, where they join the medial forebrain bundle. Together

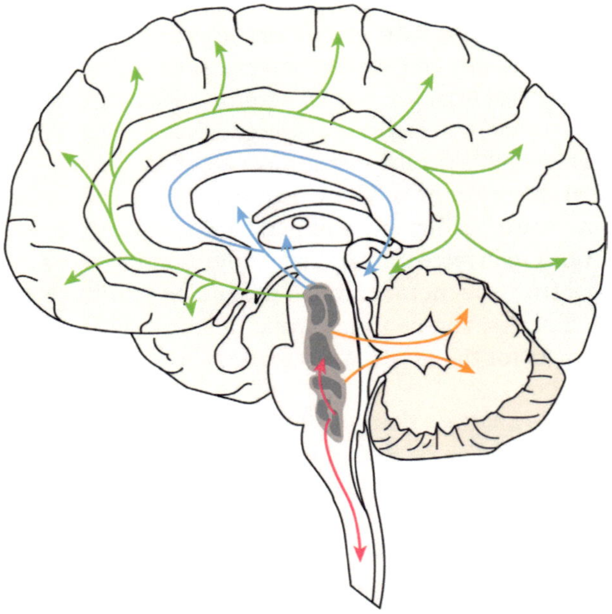

Figure 1.
Structure of the serotonergic system in the central nervous system of vertebrates. The somata of all serotonergic neurons are located within nuclei in the brainstem. These neurons extend projections to the spinal cord, cerebellum, hypothalamus, hippocampus, and throughout the cerebral cortex, thereby innervating virtually all areas of the central nervous system.

with the projections of the *locus coeruleus,* they form part of the ascending reticular activator system, which regulates attention, motor control, and sleep-wake cycles, among other functions. Ascending projections from the raphe nuclei to forebrain structures are organized in a topographical manner. The dorsal and median raphe nuclei project to forebrain regions; the median raphe projects heavily to hippocampus, *septum,* and hypothalamus, whereas the dorsal raphe innervates the striatum, and both nuclei send overlapping projections to the neocortex. Within the dorsal and median raphe, cells are organized in particular zones or groups that send axons to specific areas of the brain that are related in function. Thus, different sets of serotonergic neurons seem to be specialized for certain functions, instead of a nonspecific general innervation of the CNS [4]. In the vertebrate serotonergic system, there is reciprocal connectivity between each of the raphe nuclei and the networks it innervates [103]. All brain regions express multiple serotonin receptors, with each receptor subtype showing a specific distribution [140].

Some of the serotonergic terminals establish specialized synaptic contacts with target neurons and release serotonin upon electrical activity. However, in most of the areas in the CNS, there are at least some sites where serotonin is released without evidence of synaptic specializations [141, 142]. The axons of the serotonergic neurons from the median raphe are thick and have big spherical varicosities that form well-defined synapses in the hippocampus [143, 144]. On the other hand, axons arising from the dorsal raphe are very thin and have small spherical fusiform varicosities. In these fibers, it is difficult to demonstrate defined synaptic connections [145]. Serotonergic axons innervating the ventral horns of the spinal cord [146, 147] or the *substantia nigra reticulata* [148] establish mostly synaptic contacts with well-defined target neurons; however, in dorsal horns of the spinal cord [149] and in the *nucleus accumbens* [150], nearly 60% of the 5-HT terminals do not form synapses Recurrent axon collaterals ending in the dorsal raphe contain both synaptic and non-synaptic endings [151]. The dendrites of serotonergic neurons in the dorsal raphe nucleus also contain serotonin in small clear and large dense-core vesicles, densely packed in clusters [151–153], some of which are localized at defined synapses, but some others are not associated with synaptic structures and seem to be part of an extrasynaptic release machinery, as will be shown below.

5. Regulation of the firing rate in serotonergic neurons

Serotonergic neurons are classically thought to display a regular tonic firing at low frequencies (0.1–3 Hz), with pace-maker-like regularity [17], which is thought to maintain a tone of basal serotonin concentration in the nervous system, whereas phasic firing in bursts of higher firing rates (up to 17 Hz) [154–156] is associated with specific behaviors [157]. Serotonergic neurons in the raphe nuclei receive glutamatergic and GABAergic [158], as well as noradrenergic [159, 160] synaptic inputs that presumably contribute to the regulation of their firing frequency. In addition, serotonergic neurons of vertebrates and invertebrates connect with each other. In vertebrates, serotonergic neurons are connected through inhibitory dendro-dendritic synapses within a nucleus and by inhibitory afferents from other nuclei [93]; in invertebrates, 5-HT neurons connect through electrical synapses and through chemical synapses that are mostly excitatory, although many of the effects of serotonin are mediated by inhibitory receptors. Serotonergic neurons have 5-HT autoreceptors that participate in the regulation of their electrical activity [4, 161–166]. In rodents,

$5\text{-}HT_{1A}$ autoreceptors are localized at the somatodendritic compartment of serotonergic neurons in the raphe nuclei, whereas $5\text{-}HT_{1B}$ autoreceptors are localized at axon terminals [167, 168]. Activation of these autoreceptors activates potassium channels, producing an outward current and a robust membrane hyperpolarization [169], which decreases the firing rate of serotonergic neurons [170, 171]. On the other hand, $5\text{-}HT_2$ autoreceptors positively regulate the activity of raphe serotonergic neurons.

In Retzius neurons of the leech an autoregulation mechanism has also been found, specifically at presynaptic terminals [172], where serotonin release activates autoreceptors of unidentified type, coupled to chloride channels, that decrease the input resistance of the terminals, and their excitability [173]. Through the presence of inhibitory autoreceptors, the subsequent release of serotonin is controlled by the previous activity and release history.

The serotonin transporter, which reuptakes this amine into serotonergic neurons, also plays an important role in the regulation of their firing, by regulating the extracellular levels of serotonin and thus the autoinhibitory effect of serotonin on these neurons [174–176].

6. Serotonin secretion in the nervous system

Serotonin is released from secretory vesicles, by exocytosis. It can be stored both in small (40–60 nm) clear synaptic vesicles and in large (90–120 nm) dense-core vesicles. Each vesicle type releases its contents from different sites of the neuron, producing responses on different targets and with different time courses. Small clear vesicles generally release their contents at the active zone of presynaptic terminals, producing fast and localized effects onto specific postsynaptic terminals. In contrast, large dense-core vesicles are released at extrasynaptic sites [177–179]. Since dense-core vesicles contain 17-times more serotonin than small clear vesicles [180], exocytosis from these large vesicles releases much more serotonin, which can diffuse in the extracellular fluid and reach distant targets to produce slow and long-lasting effects over vast areas of the nervous system, in what has been called "volume transmission" [181].

6.1 Synaptic secretion

Most of the current knowledge on serotonin secretion mechanisms has been obtained from studies in identified serotonergic neurons from the central nervous system of the leech. The "colossal" Retzius neurons, so-called after the nineteenth century anatomist who described them, are the largest neurons and contain most of the serotonin in the nervous system of this invertebrate. The large size of their soma (60–100 μm in diameter) and their stereotyped localization in the ganglia facilitate their identification under simple microscopes and their electrical recordings with intracellular electrodes. In addition, these neurons can be isolated and kept in culture, where they maintain their physiological characteristics and keep synthesizing and releasing serotonin [6, 182]. In addition, if placed in contact with an adequate target neuron, they can form synapses in culture [7], which are an ideal experimental preparation to study synaptic transmission, since the isolated neurons are isopotential (changes in membrane potential are not filtered out as happens in cells with long processes) and do not receive other inputs. In these synapses, serotonin is released from small clear vesicles, which cluster near the active zone [183]. Synaptic serotonin release is quantal [6, 184], i.e., it occurs in "packs" of constant size [185], corresponding to the contents of a synaptic vesicle; it depends on calcium and the presynaptic membrane potential [8]. Like neurotransmitter release at any synapse, serotonin release at presynaptic terminals occurs within milliseconds after each action potential, since synaptic vesicles are docked at the active zone, where calcium channels are clustered [186]. Calcium entry to the terminal in response to the arrival of an action potential produces the immediate fusion of these readily releasable vesicles, as shown in **Figure 2**. Moreover, since serotonin is released from the presynaptic terminal directly to the synaptic cleft, in close proximity to receptors in the postsynaptic terminal, its effect is very fast and localized. Because presynaptic serotonergic terminals have serotonin transporters that reuptake the transmitter, synaptic effects are also short-lived, lasting only hundreds of milliseconds. Upon repetitive stimulation, synaptic transmission displays plasticity phenomena, such as synaptic facilitation [187] or depression, depending on the stimulation frequency and the release probability. In addition, serotonin-containing large dense-core vesicles are present surrounding the synaptic terminals. These large vesicles release their contents upon repetitive activity and add to the synaptic responses produced by synaptic serotonin release. Synaptic serotonin release is also regulated by autoreceptors located in the presynaptic terminals, which are linked to potassium channels (in vertebrates) or to chloride channels (in invertebrates) that lead to hyperpolarization and decrease the excitability of the terminal (**Figure 2**).

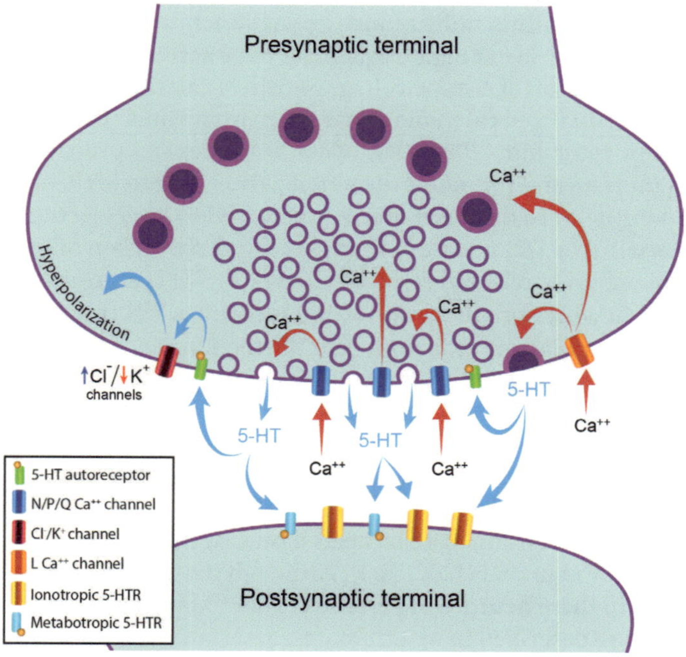

Figure 2.
Serotonin release from presynaptic terminals. Synaptic secretory vesicles are organized near or at the active zone of the presynaptic terminal. Docked vesicles can fuse with the membrane in response to single action potentials upon influx of calcium (Ca^{2+}) through voltage-gated N or P/Q types of calcium channels. Neurotransmitter released to the synaptic cleft, which is only a few nanometers wide, rapidly reaches the postsynaptic membrane and bind to its receptors. Serotonin also activates autoreceptors on the presynaptic membrane, which are coupled to potassium (K^+)or chloride (Cl^-) channels that hyperpolarize the membrane and decrease presynaptic excitability, thus immediately regulating subsequent release.

6.2 Extrasynaptic secretion

In addition to the classical mechanisms of release from presynaptic terminals, serotonin, like most neurotransmitters, is released from extrasynaptic sites of the neurons [188], including the soma, dendrites, and axonal shafts. As explained above (see Section 4.2 on the Structure of the serotonergic system), morphological evidence has shown that many of the serotonergic axonal varicosities in vertebrates contain the machinery necessary for serotonin release, but do not have postsynaptic counterparts [141, 142, 145, 149–151], suggesting that these sites display extrasynaptic secretion. In addition, evidence from microdialysis and cyclic voltammetry showed the presence of serotonin in the extracellular fluid in the CNS of vertebrates and invertebrates [189–195], in concentration that match the affinity of 5-HT receptors [189]. Moreover, there is extensive evidence for the existence of receptors and transporters for serotonin at extrasynaptic locations [141, 168, 175, 196–198].

Extrasynaptic secretion has characteristics that differ a lot from those of synaptic secretion and are more similar to hormone secretion by neuroendocrine cells. The first direct demonstration of extrasynaptic serotonin secretion was made in the soma of leech Retzius neurons [199]. In these neurons, in contrast to presynaptic terminals, where vesicles are docked at the presynaptic active zone, very near calcium channels, the soma has clusters of hundreds of large dense-core vesicles, forming different pools that rest at different distances from the plasma membrane. This difference in the localization of vesicles at rest produces large differences in the requirements for the activation of somatic secretion and in its time course. In contrast to synaptic secretion, which is activated within milliseconds by single action potentials, somatic secretion requires repetitive firing at high frequencies to be activated. Somatic secretion requires also the activation of L-type voltage-sensitive calcium channels [199] and depends on calcium-induced calcium release from intracellular calcium pools, such as the endoplasmic reticulum [200, 201], which amplifies and propagates the calcium signal in the cytoplasm and promotes the mobilization of vesicle clusters (**Figure 3**). This confers a long delay to the onset of somatic secretion after electrical activity, which contrasts with the fast release at synapses. Upon the fusion of the first vesicles that reach the membrane, released serotonin activates 5-HT_2 autoreceptors coupled to phospholipase C, which produces the second messenger IP3 that in turn releases calcium from the endoplasmic reticulum near the membrane. This calcium promotes the fusion of vesicles subsequently reaching the membrane, establishing a positive feedback cycle that sustains serotonin release for minutes after only a short burst of action potentials at high frequency, until all the vesicles mobilized by the first calcium signal release their contents [200]. The mechanism producing somatic serotonin secretion is illustrated in **Figure 3**.

Somatic release of serotonin has also been shown by multi-photon microscopy in the raphe nuclei of mammals [202, 203]. Although the detailed mechanisms of somatic secretion in these neurons in vertebrates have not been elucidated, they seem to be similar to those described in invertebrates, and somatic release lasts also for minutes, reaching vast volumes in the nervous system [204].

In addition to somatic secretion, serotonergic neurons have been shown to display extrasynaptic secretion from the dendrites. Dendritic serotonin release can be activated by glutamatergic inputs, which open L-type calcium channels in the dendrites, independently of action potential firing [205].

The presence of clusters of secretory vesicles in different sites of serotonergic axons in the absence of postsynaptic counterparts, as described above, strongly

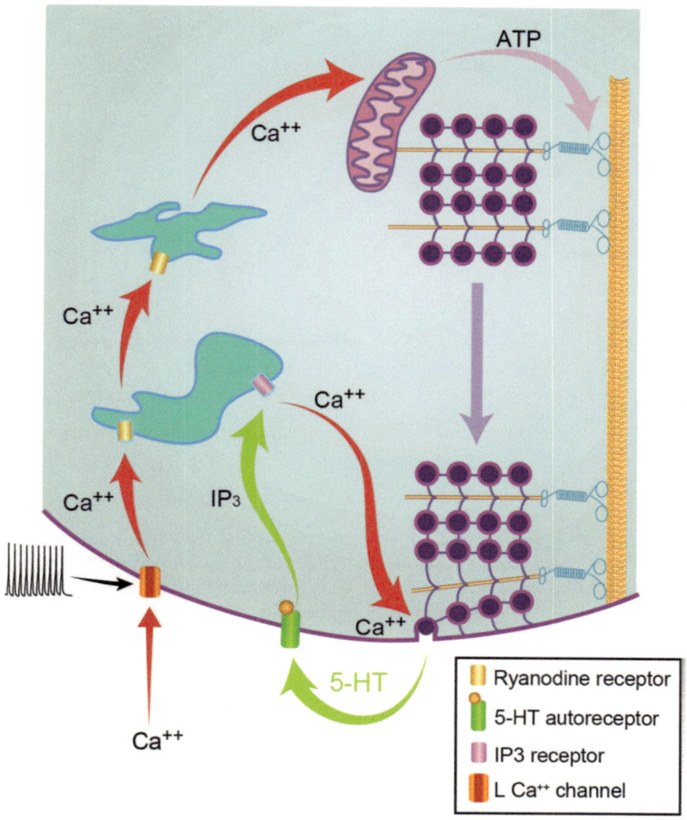

Figure 3.
Schematic diagram of somatic secretion of serotonin (5-HT) from leech Retzius neurons. Clusters of dense-core vesicles distant from the plasma membrane are mobilized by motor proteins activated by ATP and by Ca^{2+} waves initiated by entry through L-type channels and propagated by calcium-induced calcium release. A 5-HT-activated feedback loop mediated by IP3-induced intracellular Ca^{2+} release sustains secretion in these cells. All of these mechanisms are also seen in vertebrates.

suggests that extrasynaptic serotonin release takes place also from the axons. Extrasynaptic secretion of serotonin from the primary axon has been directly observed in leech Retzius neurons, where the requirements and characteristics of secretion are intermediate between those of release at presynaptic terminals and that of somatic secretion (Cercós and Trueta, in preparation). Extrasynaptic release from sites along the axons of serotonergic neurons could act in a more diffuse manner than serotonin released from presynaptic terminals, but produce effects still somewhat restricted in space, thus producing local neuromodulation at different levels of neural circuits.

In addition, in vertebrates and invertebrates, serotonin is secreted from neurosecretory organs to the circulatory system, through which it can reach peripheral organs and act as a neurohormone. Thus, serotonin can act in at least three ways: as a neurotransmitter at specialized synapses, a neuromodulator at extrasynaptic paracrine secretion sites, and a neurohormone in the general circulation.

Due to the differences in the localization of the vesicles that produce secretion in each neuronal compartment, the different modes of serotonin secretion in the nervous system have very different time courses. Synaptic release occurs upon each action potential, and is thus synchronized with electrical activity, and its effects are

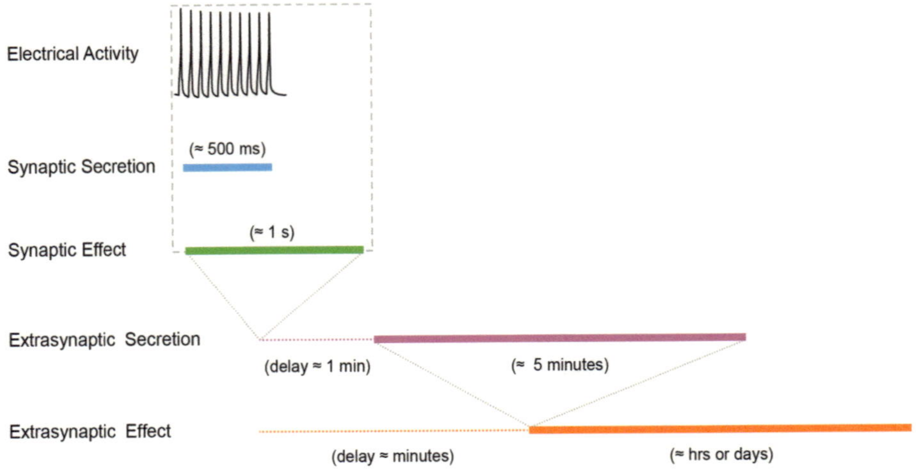

Figure 4.
Differences in the time courses of synaptic and extrasynaptic release and their effects on the nervous system. In response to a burst of action potentials lasting 500 ms, release at synaptic terminals occurs synchronized with electrical activity and produces postsynaptic effects lasting a few hundred of milliseconds more. In contrast, extrasynaptic release begins with a delay of nearly a minute and lasts several minutes and its effects, mediated by diffusion, may last for hours or even longer.

produced immediately, lasting only hundreds of milliseconds. In contrast, extrasynaptic release occurs with a long delay after electrical activity, and its effects are even more delayed, due to the diffusion of the transmitter in the extracellular space before reaching its targets. Extrasynaptic secretion also lasts for minutes after a short train of action potential [199, 200], and its effects may last for hours or even longer periods. **Figure 4** schematizes the differences in the time courses of synaptic and extrasynaptic secretion and their effects on the nervous system.

Neurotransmitter release at synapses has immediate and localized effects that contribute to synaptic computation in neuronal networks. On the other hand, extrasynaptic release from the soma, axons, or dendrite is thought to produce slow and diffuse neuromodulatory effects on neuronal populations, changing the way in which the synaptic networks respond. These slow effects may be related to the regulation of mood, emotions, and social behavior, which have time courses that exceed by far the effects of synaptic transmission. However, there is no concrete evidence of the effects of extrasynaptic release for serotonin or other neurotransmitters in the nervous system. This is a fascinating subject for further research.

The different modes of serotonin release, from different compartments of the same neurons, enable these neurons to act as multifunctional cells. This may be the way in which these neurons, although reduced in number, are able to regulate such a wide diversity of functions in the nervous system.

7. Conclusions

Serotonin in the nervous system is synthesized by small numbers of neurons that release it by exocytosis from different compartments of their complex structure. The type of structure where serotonin is released defines the type of information processing that serotonin carries in each region. Presynaptic terminals release small amounts of serotonin, in synchrony with electrical activity, and with a strong association between the presynaptic neuron and its target, producing very fast, local, and short-lasting effects on specific postsynaptic targets that participate in immediate functions. On the other hand, extrasynaptic sites of these neurons, including the soma and the axons, slowly release massive amounts of serotonin that produce less localized and more dynamic interactions with the target neurons, whereby serotonin can reach distant and diverse targets by diffusion and produce slow and long-lasting neuromodulatory effects, such as those that characterize emotions and social behavior. By displaying different modes of neurotransmitter release, small numbers of serotonergic neurons can regulate a multiplicity of functions in the nervous system, which have been conserved throughout the phylogenetic scale.

Acknowledgements

We thank Ricardo Cruz Garduño for the elaboration of the figures for this chapter, and Silvia Rocío Mejía Mauríes for technical and administrative assistance.

This work has been funded by INPRFM internal resources. Dr. Nuria Segovia paid the publishing fee. We are thankful for her generosity.

Conflict of interest

The authors declare no conflict of interest.

Author details

Citlali Trueta* and Montserrat G. Cercós
Departamento de Neurofisiología, Instituto Nacional de Psiquiatría Ramón de la Fuente Muñiz, México City, Mexico

*Address all correspondence to: ctrueta@inprf.gob.mx

References

[1] Erspamer V, Vialli M. Ricerche sul secreto delle cellule enterocromaffini. Bolletino da Società Medico-Chirurgico di Pavia. 1937;**51**:357-363

[2] Whitaker-Azmitia P. The discovery of serotonin and its role in neuroscience. Neuropsychopharmacology. 1999;**21**(2):2S-8S

[3] Csaba G. Presence in and effects of pineal indoleamines at very low level of phylogeny. Experientia. 1993;**49**(8):627-634

[4] Frazer A, Hensler JG. Serotonin. In: Siegel GARW, Brady SPD, editors. Basic Neurochemistry Molecular, Cellular, and Medical Aspects. 7th ed. Amsterdam, Netherlands: Elsevier; 2006

[5] Bruns D, Engert F, Lux HD. A fast activating presynaptic reuptake current during serotonergic transmission in identified neurons of Hirudo. Neuron. 1993;**10**(4):559-572

[6] Henderson LP, Kuffler DP, Nicholls J, Zhang R. Structural and functional analysis of synaptic transmission between identified leech neurones in culture. Journal of Physiology (London). 1983;**340**:347-358

[7] Fuchs PA, Henderson LP, Nicholls JG. Chemical transmission between individual Retzius and sensory neurones of the leech in culture. Journal of Physiology (London). 1982;**323**:195-210

[8] Dietzel ID, Drapeau P, Nicholls JG. Voltage dependence of 5-hydroxytryptamine release at a synapse between identified leech neurones in culture. Journal of Physiology (London). 1986;**372**:191-205

[9] Chattopadhyay AA, editor. Serotonin Receptors in Neurobiology. Boca Raton (FL): CRC Press; 2007 (Frontiers in Neuroscience)

[10] Göthert M. Presynaptic serotonin receptors in the central nervous system. Annals of the New York Academy of Sciences. 1990;**604**:102-112

[11] Kroeze WK, Kristiansen K, Roth BL. Molecular biology of serotonin receptors structure and function at the molecular level. Current Topics in Medicinal Chemistry. 2002;**2**(6):507-528

[12] Nichols DE, Nichols CD. Serotonin receptors. Chemical Reviews. 2008;**108**(5):1614-1641

[13] Palacios JM, Waeber C, Hoyer D, Mengod G. Distribution of serotonin receptors. Annals of the New York Academy of Sciences. 1990;**600**:36-52

[14] Jacobs BL, Fornal CA. 5-HT and motor control: A hypothesis. Trends in Neurosciences. 1993;**16**(9):346-352

[15] McCall RB, Aghajanian GK. Serotonergic facilitation of facial motoneuron excitation. Brain Research. 1979;**169**(1):11-27

[16] White SR, Fung SJ, Jackson DA, Imel KM. Serotonin, norepinephrine and associated neuropeptides: Effects on somatic motoneuron excitability. Progress in Brain Research. 1996;**107**:183-199

[17] Jacobs BL, Azmitia EC. Structure and function of the brain serotonin system. Physiological Reviews. 1992;**72**(1):165-229

[18] Prosser RA, Miller JD, Heller HC. A serotonin agonist phase-shifts the

circadian clock in the suprachiasmatic nuclei in vitro. Brain Research. 1990;**534**(1-2):336-339

[19] Aghajanian GK, Sanders-Bush E. Serotonin. In: Davis KL, Charney D, Coyle JT, Nemeroff C, editors. Neuropsychopharmacology - The Fifth Generation of Progress [Internet]. New York: Lippincott, Williams & Wilkins; 2002. pp. 3-23. Available from: https://acnp.org/wp-content/uploads/2017/11/CH2_15-34.pdf

[20] Kravitz EA. Serotonin and aggression: Insights gained from a lobster model system and speculations on the role of amine neurons in a complex behavior. Journal of Comparative Physiology. A. 2000;**186**(3):221-238

[21] Arango V, Underwood MD, Mann JJ. Serotonin brain circuits involved in major depression and suicide. Progress in Brain Research. 2002;**136**:443-453

[22] Brieden T, Ujeyl M, Naber D. Psychopharmacological treatment of aggression in schizophrenic patients. Pharmacopsychiatry. 2002;**35**(3):83-89

[23] Roth BL, Hanizavareh SM, Blum AE. Serotonin receptors represent highly favorable molecular targets for cognitive enhancement in schizophrenia and other disorders. Psychopharmacology. 2004;**174**(1):17-24

[24] Jenike MA, Rauch SL, Cummings JL, Savage CR, Goodman WK. Recent developments in neurobiology of obsessive-compulsive disorder. The Journal of Clinical Psychiatry. 1996;**57**(10):492-503

[25] Kaye W. Neurobiology of anorexia and bulimia nervosa. Physiology & Behavior. 2008;**94**(1):121-135

[26] Mar A, Drapeau P. Modulation of conduction block in leech mechanosensory neurons. The Journal of Neuroscience. 1996;**16**(14):4335-4343

[27] Nicholls JG, Baylor DA. Specific modalities and receptive fields of sensory neurons in CNS of the leech. Journal of Neurophysiology. 1968;**31**(5):740-756

[28] Friesen WO, Poon M, Stent GS. An oscillatory neuronal circuit generating a locomotory rhythm. Proceedings of the National Academy of Sciences of the United States of America. 1976;**73**(10):3734-3738

[29] Nusbaum MP, Kristan WB. Swim initiation in the leech by serotonin-containing interneurones, cells 21 and 61. The Journal of Experimental Biology. 1986;**122**:277-302

[30] Nusbaum MP. Synaptic basis of swim initiation in the leech. III. Synaptic effects of serotonin-containing interneurones (cells 21 and 61) on swim CPG neurones (cells 18 and 208). The Journal of Experimental Biology. 1986;**122**:303-321

[31] Willard AL. Effects of serotonin on the generation of the motor program for swimming by the medicinal leech. The Journal of Neuroscience. 1981;**1**(9):936-944

[32] Angstadt JD, Friesen WO. Modulation of swimming behavior in the medicinal leech. I. Effects of serotonin on the electrical properties of swim-gating cell 204. Journal of Comparative Physiology. A. 1993;**172**(2):223-234

[33] Angstadt JD, Friesen WO. Modulation of swimming behavior in the medicinal leech. II. Ionic conductances underlying serotonergic modulation of swim-gating cell 204. Journal of Comparative Physiology. A. 1993;**172**(2):235-248

[34] Mangan PS, Cometa AK, Friesen WO. Modulation of swimming

behavior in the medicinal leech. IV. Serotonin-induced alteration of synaptic interactions between neurons of the swim circuit. Journal of Comparative Physiology. A. 1994;**175**(6):723-736

[35] Mangan PS, Curran GA, Hurney CA, Friesen WO. Modulation of swimming behavior in the medicinal leech. III. Control of cellular properties in motor neurons by serotonin. Journal of Comparative Physiology. A. 1994;**175**(6):709-722

[36] Lent CM, Dickinson MH. Retzius cells retain functional membrane properties following "ablation" by the neurotoxin 5,7-DHT. Brain Research. 1984;**300**(1):167-171

[37] Okusawa S, Kohsaka H, Nose A. Serotonin and downstream leucokinin neurons modulate larval turning behavior in Drosophila. The Journal of Neuroscience. 2014;**34**(7):2544-2558

[38] Majeed ZR, Abdeljaber E, Soveland R, Cornwell K, Bankemper A, Koch F, et al. Modulatory action by the serotonergic system: Behavior and neurophysiology in *Drosophila melanogaster*. Neural Plasticity. 2016;**2016**:7291438

[39] Wilson RJ, Skierczynski BA, Blackwood S, Skalak R, Kristan WB. Mapping motor neurone activity to overt behaviour in the leech: Internal pressures produced during locomotion. The Journal of Experimental Biology. 1996;**199**(Pt 6):1415-1428

[40] Sawada M, Coggeshall RE. Ionic mechanism of 5-hydroxytryptamine induced hyperpolarization and inhibitory junctional potential in body wall muscle cells of *Hirudo medicinalis*. Journal of Neurobiology. 1976;7(1):63-73

[41] Sawada M, Coggeshall RE. A central inhibitory action of 5-hydroxytryptamine in the leech. Journal of Neurobiology. 1976;7(6):477-482

[42] Lent CM. Retzius cells: Neuroeffectors controlling mucus release by the leech. Science. 1973;**179**(4074):693-696

[43] Groome JR, Clark M, Lent CM. The behavioural state of satiation in the leech is regulated by body distension and mimicked by serotonin depletion. The Journal of Experimental Biology. 1993;**182**:265-270

[44] Lent CM. Serotonergic modulation of the feeding behavior of the medicinal leech. Brain Research Bulletin. 1985;**14**(6):643-655

[45] Ehrlich JS, Boulis NM, Karrer T, Sahley CL. Differential effects of serotonin depletion on sensitization and dishabituation in the leech, *Hirudo medicinalis*. Journal of Neurobiology. 1992;**23**(3):270-279

[46] Sahley CL. Serotonin depletion impairs but does not eliminate classical conditioning in the leech *Hirudo medicinalis*. Behavioral Neuroscience. 1994;**108**(6):1043-1052

[47] Burrell BD, Sahley CL, Muller KJ. Non-associative learning and serotonin induce similar bi-directional changes in excitability of a neuron critical for learning in the medicinal leech. The Journal of Neuroscience. 2001;**21**(4):1401-1412

[48] Burrell BD, Sahley CL, Muller KJ. Differential effects of serotonin enhance activity of an electrically coupled neural network. Journal of Neurophysiology. 2002;**87**(6):2889-2895

[49] Satterlie RA, Norekian TP. Modulation of swimming speed in the

pteropod mollusc, Clione limacina: Role of a compartmental serotonergic system. Invertebrate Neuroscience. 1996;**2**(3):157-165

[50] Castellucci V, Kandel ER. Presynaptic facilitation as a mechanism for behavioral sensitization in Aplysia. Science. 1976;**194**(4270):1176-1178

[51] Marinesco S, Carew TJ. Serotonin release evoked by tail nerve stimulation in the CNS of aplysia: Characterization and relationship to heterosynaptic plasticity. The Journal of Neuroscience. 2002;**22**(6):2299-2312

[52] Levenson J, Byrne JH, Eskin A. Levels of serotonin in the hemolymph of Aplysia are modulated by light/dark cycles and sensitization training. The Journal of Neuroscience. 1999;**19**(18):8094-8103

[53] Siegelbaum SA, Camardo JS, Kandel ER. Serotonin and cyclic AMP close single K+ channels in Aplysia sensory neurones. Nature. 1982;**299**(5882):413-417

[54] Baxter DA, Byrne JH. Serotonergic modulation of two potassium currents in the pleural sensory neurons of Aplysia. Journal of Neurophysiology. 1989;**62**(3):665-679

[55] Klein M, Camardo J, Kandel ER. Serotonin modulates a specific potassium current in the sensory neurons that show presynaptic facilitation in Aplysia. Proceedings of the National Academy of Sciences of the United States of America. 1982;**79**(18):5713-5717

[56] Braha O, Dale N, Hochner B, Klein M, Abrams TW, Kandel ER. Second messengers involved in the two processes of presynaptic facilitation that contribute to sensitization and dishabituation in Aplysia sensory neurons. Proceedings of the National Academy of Sciences of the United States of America. 1990;**87**(5):2040-2044

[57] Hochner B, Klein M, Schacher S, Kandel ER. Additional component in the cellular mechanism of presynaptic facilitation contributes to behavioral dishabituation in Aplysia. Proceedings of the National Academy of Sciences of the United States of America. 1986;**83**(22):8794-8798

[58] Hochner B, Klein M, Schacher S, Kandel ER. Action-potential duration and the modulation of transmitter release from the sensory neurons of Aplysia in presynaptic facilitation and behavioral sensitization. Proceedings of the National Academy of Sciences of the United States of America. 1986;**83**(21):8410-8414

[59] Klein M. Differential cyclic AMP dependence of facilitation at Aplysia sensorimotor synapses as a function of prior stimulation: Augmentation versus restoration of transmitter release. The Journal of Neuroscience. 1993;**13**(9):3793-3801

[60] Klein M. Synaptic augmentation by 5-HT at rested Aplysia sensorimotor synapses: Independence of action potential prolongation. Neuron. 1994;**13**(1):159-166

[61] Sugita S, Goldsmith JR, Baxter DA, Byrne JH. Involvement of protein kinase C in serotonin-induced spike broadening and synaptic facilitation in sensorimotor connections of Aplysia. Journal of Neurophysiology. 1992;**68**(2):643-651

[62] Barbas D, DesGroseillers L, Castellucci VF, Carew TJ, Marinesco S. Multiple serotonergic mechanisms contributing to sensitization in aplysia: Evidence of diverse serotonin receptor subtypes. Learning & Memory. 2003;**10**(5):373-386

[63] Livingstone MS, Harris-Warrick RM, Kravitz EA. Serotonin and octopamine produce opposite postures in lobsters. Science. 1980;**208**(4439):76-79

[64] Harris-Warrick RM, Kravitz EA. Cellular mechanisms for modulation of posture by octopamine and serotonin in the lobster. The Journal of Neuroscience. 1984;**4**(8):1976-1993

[65] Dixon D, Atwood HL. Phosphatidylinositol system's role in serotonin-induced facilitation at the crayfish neuromuscular junction. Journal of Neurophysiology. 1989;**62**(1):239-246

[66] Glusman S, Kravitz EA. The action of serotonin on excitatory nerve terminals in lobster nerve-muscle preparations. The Journal of Physiology. 1982;**325**:223-241

[67] Goy MF, Kravitz EA. Cyclic AMP only partially mediates the actions of serotonin at lobster neuromuscular junctions. The Journal of Neuroscience. 1989;**9**(1):369-379

[68] Kravitz EA. Hormonal control of behavior: Amines and the biasing of behavioral output in lobsters. Science. 1988;**241**(4874):1775-1781

[69] Chen S, Lee AY, Bowens NM, Huber R, Kravitz EA. Fighting fruit flies: A model system for the study of aggression. Proceedings of the National Academy of Sciences of the United States of America. 2002;**99**(8):5664-5668

[70] Yurkovic A, Wang O, Basu AC, Kravitz EA. Learning and memory associated with aggression in *Drosophila melanoga*ster. Proceedings of the National Academy of Sciences of the United States of America. 2006;**103**(46):17519-17524

[71] Dierick HA, Greenspan RJ. Serotonin and neuropeptide F have opposite modulatory effects on fly aggression. Nature Genetics. 2007;**39**(5):678-682

[72] Alekseyenko OV, Lee C, Kravitz EA. Targeted manipulation of serotonergic neurotransmission affects the escalation of aggression in adult male *Drosophila melanogaster*. PLoS One. 2010:e10806

[73] Artiushin G, Sehgal A. The Drosophila circuitry of sleep-wake regulation. Current Opinion in Neurobiology. 2017;**44**:243-250

[74] Strauss J, Dircksen H. Circadian clocks in crustaceans: Identified neuronal and cellular systems. Frontiers in Bioscience (Landmark edition). 2010;**15**(3):1040-1074

[75] Escamilla-Chimal EG, Van Herp F, Fanjul-Moles ML. Daily variations in crustacean hyperglycaemic hormone and serotonin immunoreactivity during the development of crayfish. The Journal of Experimental Biology. 2001;**204**(Pt 6):1073-1081

[76] Harris-Warrick RM, Cohen AH. Serotonin modulates the central pattern generator for locomotion in the isolated lamprey spinal cord. The Journal of Experimental Biology. 1985;**116**:27-46

[77] Buchanan JT. Flexibility in the patterning and control of axial locomotor networks in lamprey. Integrative and Comparative Biology. 2011;**51**(6):869-878

[78] Wallén P, Christenson J, Brodin L, Hill R, Lansner A, Grillner S. Mechanisms underlying the serotonergic modulation of the spinal circuitry for locomotion in lamprey. Progress in Brain Research. 1989;**80**:321-327. discussion 315-319

[79] Daghfous G, Green WW, Alford ST, Zielinski BS, Dubuc R. Sensory activation of command cells for

locomotion and modulatory mechanisms: Lessons from lampreys. Frontiers in Neural Circuits. 2016;**10**:18

[80] Lima-Maximino M, Pyterson MP, do Carmo Silva RX, GCV G, Rocha SP, Herculano AM, et al. Phasic and tonic serotonin modulate alarm reactions and post-exposure behavior in zebrafish. Journal of Neurochemistry. 2020;**153**(4):495-509

[81] Brustein E, Chong M, Holmqvist B, Drapeau P. Serotonin patterns locomotor network activity in the developing zebrafish by modulating quiescent periods. Journal of Neurobiology. 2003;**57**(3):303-322

[82] Gabriel JP, Mahmood R, Kyriakatos A, Söll I, Hauptmann G, Calabrese RL, et al. Serotonergic modulation of locomotion in zebrafish: Endogenous release and synaptic mechanisms. The Journal of Neuroscience. 2009;**29**(33):10387-10395

[83] Montgomery JE, Wahlstrom-Helgren S, Wiggin TD, Corwin BM, Lillesaar C, Masino MA. Intraspinal serotonergic signaling suppresses locomotor activity in larval zebrafish. Developmental Neurobiology. 2018;**78**:807-827

[84] Hachoumi L, Sillar KT. Developmental stage-dependent switching in the neuromodulation of vertebrate locomotor central pattern generator networks. Developmental Neurobiology. 2020;**80**(1-2):42-57

[85] McLean DL, Merrywest SD, Sillar KT. The development of neuromodulatory systems and the maturation of motor patterns in amphibian tadpoles. Brain Research Bulletin. 2000;**53**(5):595-603

[86] Cabaj AM, Majczyński H, Couto E, Gardiner PF, Stecina K, Sławińska U, et al. Serotonin controls initiation of locomotion and afferent modulation of coordination via 5-HT7 receptors in adult rats. The Journal of Physiology. 2017;**595**(1):301-320

[87] Uchida N, Cohen JY. Slow motion. Elife. 2017;**6**:e24792

[88] Viala D, Buser P. Effects of a decarboxylase inhibitor on the Dopa and 5-HTP induced changes in the locomotor-like discharge pattern of rabbit hind limb nerves. Psychopharmacologia. 1974;**40**(3):225-233

[89] Sławińska U, Miazga K, Jordan LM. The role of serotonin in the control of locomotor movements and strategies for restoring locomotion after spinal cord injury. Acta Neurobiologiae Experimentalis (Wars). 2014;**74**(2):172-187

[90] Liu J, Jordan LM. Stimulation of the parapyramidal region of the neonatal rat brain stem produces locomotor-like activity involving spinal 5-HT7 and 5-HT2A receptors. Journal of Neurophysiology. 2005;**94**(2):1392-1404

[91] Norreel JC, Pflieger JF, Pearlstein E, Simeoni-Alias J, Clarac F, Vinay L. Reversible disorganization of the locomotor pattern after neonatal spinal cord transection in the rat. The Journal of Neuroscience. 2003;**23**(5):1924-1932

[92] Oueghlani Z, Juvin L, Lambert FM, Cardoit L, Courtand G, Masmejean F, et al. Serotonergic modulation of sacral dorsal root stimulation-induced locomotor output in newborn rat. Neuropharmacology. 2020;**170**:107815

[93] Gillette R. Evolution and function in serotonergic systems. Integrative and Comparative Biology. 2006;**46**(6):838-846

[94] Heisler LK, Cowley MA, Tecott LH, Fan W, Low MJ, Smart JL, et al. Activation of central melanocortin pathways by fenfluramine. Science. 2002;**297**(5581):609-611

[95] Sohn JW, Xu Y, Jones JE, Wickman K, Williams KW, Elmquist JK. Serotonin 2C receptor activates a distinct population of arcuate pro-opiomelanocortin neurons via TRPC channels. Neuron. 2011;**71**(3):488-497

[96] Heisler LK, Jobst EE, Sutton GM, Zhou L, Borok E, Thornton-Jones Z, et al. Serotonin reciprocally regulates melanocortin neurons to modulate food intake. Neuron. 2006;**51**(2):239-249

[97] Breisch ST, Zemlan FP, Hoebel BG. Hyperphagia and obesity following serotonin depletion by intraventricular p-chlorophenylalanine. Science. 1976;**192**(4237):382-385

[98] Saller CF, Stricker EM. Hyperphagia and increased growth in rats after intraventricular injection of 5,7-dihydroxytryptamine. Science. 1976;**192**(4237):385-387

[99] Bouwknecht JA, van der Gugten J, Hijzen TH, Maes RA, Hen R, Olivier B. Male and female 5-HT(1B) receptor knockout mice have higher body weights than wildtypes. Physiology & Behavior. 2001;**74**(4-5):507-516

[100] Tecott LH, Sun LM, Akana SF, Strack AM, Lowenstein DH, Dallman MF, et al. Eating disorder and epilepsy in mice lacking 5-HT2c serotonin receptors. Nature. 1995;**374**(6522):542-546

[101] Dryden S, Frankish HM, Wang Q, Pickavance L, Williams G. The serotonergic agent fluoxetine reduces neuropeptide Y levels and neuropeptide Y secretion in the hypothalamus of lean and obese rats. Neuroscience. 1996;**72**(2):557-566

[102] Xu Y, Jones JE, Kohno D, Williams KW, Lee CE, Choi MJ, et al. 5-HT2CRs expressed by pro-opiomelanocortin neurons regulate energy homeostasis. Neuron. 2008;**60**(4):582-589

[103] Jacobs BL, Fornal CA. Activity of serotonergic neurons in behaving animals. Neuropsychopharmacology. 1999;**21**(Suppl. 2):9S-15S

[104] Monti JM. Serotonin control of sleep-wake behavior. Sleep Medicine Reviews. 2011;**15**(4):269-281

[105] Portas CM, Bjorvatn B, Ursin R. Serotonin and the sleep/wake cycle: Special emphasis on microdialysis studies. Progress in Neurobiology. 2000;**60**(1):13-35

[106] Morin LP. Serotonin and the regulation of mammalian circadian rhythmicity. Annals of Medicine. 1999;**31**(1):12-33

[107] Paredes SD, Terrón MP, Cubero J, Valero V, Barriga C, Reiter RJ, et al. Comparative study of the activity/rest rhythms in young and old ringdove (*Streptopelia risoria*): Correlation with serum levels of melatonin and serotonin. Chronobiology International. 2006;**23**(4):779-793

[108] Ferris CF. Adolescent stress and neural plasticity in hamsters: A vasopressin-serotonin model of inappropriate aggressive behaviour. Experimental Physiology. 2000;**85 Spec No**:85S-90S

[109] Weiger WA. Serotonergic modulation of behaviour: A phylogenetic overview. Biological Reviews of the Cambridge Philosophical Society. 1997;**72**(1):61-95

[110] Miczek KA, de Almeida RMM, Kravitz EA, Rissman EF, de Boer SF, Raine A. Neurobiology of escalated aggression and violence. Journal of Neuroscience. 2007;**27**(44):11803-11806

[111] Brunner D, Hen R. Insights into the neurobiology of impulsive behavior from serotonin receptor knockout mice. Annals of the New York Academy of Sciences. 1997;**836**:81-105

[112] de Boer SF, van der Vegt BJ, Koolhaas JM. Individual variation in aggression of feral rodent strains: A standard for the genetics of aggression and violence? Behavior Genetics. 2003;**33**(5):485-501

[113] de Boer SF, Caramaschi D, Natarajan D, Koolhaas JM. The vicious cycle towards violence: Focus on the negative feedback mechanisms of brain serotonin neurotransmission. Frontiers in Behavioral Neuroscience. 2009;**52**:1-6. Article 52

[114] Ferrari PF, van Erp AMM, Tornatzky W, Miczek KA. Accumbal dopamine and serotonin in anticipation of the next aggressive episode in rats. The European Journal of Neuroscience. 2003;**17**(2):371-378

[115] van Erp AM, Miczek KA. Aggressive behavior, increased accumbal dopamine, and decreased cortical serotonin in rats. The Journal of Neuroscience. 2000;**20**(24):9320-9325

[116] Coccaro EF. Central serotonin and impulsive aggression. The British Journal of Psychiatry. Supplement. 1989;**8**:52-62

[117] Linnoila M, Virkkunen M, Scheinin M, Nuutila A, Rimon R, Goodwin FK. Low cerebrospinal fluid 5-hydroxyindoleacetic acid concentration differentiates impulsive from nonimpulsive violent behavior. Life Sciences. 1983;**33**(26):2609-2614

[118] van der Vegt BJ, Lieuwes N, Cremers TIFH, de Boer SF, Koolhaas JM. Cerebrospinal fluid monoamine and metabolite concentrations and aggression in rats. Hormones and Behavior. 2003;**44**(3):199-208

[119] van der Vegt BJ, Lieuwes N, van de Wall EHEM, Kato K, Moya-Albiol L, Martínez-Sanchis S, et al. Activation of serotonergic neurotransmission during the performance of aggressive behavior in rats. Behavioral Neuroscience. 2003;**117**(4):667-674

[120] Higley JD, Linnoila M. Low central nervous system serotonergic activity is traitlike and correlates with impulsive behavior. A nonhuman primate model investigating genetic and environmental influences on neurotransmission. Annals of the New York Academy of Sciences. 1997;**836**:39-56

[121] Raleigh MJ, McGuire MT, Brammer GL, Pollack DB, Yuwiler A. Serotonergic mechanisms promote dominance acquisition in adult male vervet monkeys. Brain Research. 1991;**559**(2):181-190

[122] Vadodaria KC, Stern S, Marchetto MC, Gage FH. Serotonin in psychiatry: In vitro disease modeling using patient-derived neurons. Cell and Tissue Research. 2018;**371**(1):161-170

[123] Jobe PC, Dailey JW, Wernicke JF. A noradrenergic and serotonergic hypothesis of the linkage between epilepsy and affective disorders. Critical Reviews in Neurobiology. 1999;**13**(4):317-356

[124] Mann JJ, McBride PA, Brown RP, Linnoila M, Leon AC, DeMeo M, et al. Relationship between central and

peripheral serotonin indexes in depressed and suicidal psychiatric inpatients. Archives of General Psychiatry. 1992;**49**(6):442-446

[125] Asberg M, Träskman L, Thorén P. 5-HIAA in the cerebrospinal fluid. A biochemical suicide predictor? Archives of General Psychiatry. 1976;**33**(10):1193-1197

[126] Krishnan V, Nestler EJ. The molecular neurobiology of depression. Nature. 2008;**455**(7215):894-902

[127] Mahar I, Bambico FR, Mechawar N, Nobrega JN. Stress, serotonin, and hippocampal neurogenesis in relation to depression and antidepressant effects. Neuroscience and Biobehavioral Reviews. 2014;**38**:173-192

[128] Grahn RE, Will MJ, Hammack SE, Maswood S, McQueen MB, Watkins LR, et al. Activation of serotonin-immunoreactive cells in the dorsal raphe nucleus in rats exposed to an uncontrollable stressor. Brain Research. 1999;**826**(1):35-43

[129] Graeff FG, Guimarães FS, De Andrade TG, Deakin JF. Role of 5-HT in stress, anxiety, and depression. Pharmacology, Biochemistry, and Behavior. 1996;**54**(1):129-141

[130] Zangrossi H, Viana MB, Zanoveli J, Bueno C, Nogueira RL, Graeff FG. Serotonergic regulation of inhibitory avoidance and one-way escape in the rat elevated T-maze. Neuroscience and Biobehavioral Reviews. 2001;**25**(7-8):637-645

[131] Gross C, Santarelli L, Brunner D, Zhuang X, Hen R. Altered fear circuits in 5-HT(1A) receptor KO mice. Biological Psychiatry. 2000;**48**(12):1157-1163

[132] Albert PR, Vahid-Ansari F, Luckhart C. Serotonin-prefrontal cortical circuitry in anxiety and depression phenotypes: Pivotal role of pre- and post-synaptic 5-HT1A receptor expression. Frontiers in Behavioral Neuroscience. 2014;**8**:199

[133] Hull EM, Lorrain DS, Du J, Matuszewich L, Lumley LA, Putnam SK, et al. Hormone-neurotransmitter interactions in the control of sexual behavior. Behavioural Brain Research. 1999;**105**(1):105-116

[134] Wang QP, Nakai Y. The dorsal raphe: An important nucleus in pain modulation. Brain Research Bulletin. 1994;**34**(6):575-585

[135] Remington G. Alterations of dopamine and serotonin transmission in schizophrenia. Progress in Brain Research. 2008;**172**:117-140

[136] Humphrey PP, Feniuk W, Perren MJ, Beresford IJ, Skingle M, Whalley ET. Serotonin and migraine. Annals of the New York Academy of Sciences. 1990;**600**:587-600; discussion 598-600

[137] Underwood MD, Khaibulina AA, Ellis SP, Moran A, Rice PM, Mann JJ, et al. Morphometry of the dorsal raphe nucleus serotonergic neurons in suicide victims. Biological Psychiatry. 1999;**46**(4):473-483

[138] Lent CM, Frazer BM. Connectivity of the monoamine-containing neurones in central nervous system of leech. Nature. 1977;**266**(5605):844-847

[139] Dahlström A, Fuxe K. Localization of monoamines in the lower brain stem. Experientia. 1964;**20**(7):398-399

[140] Mengold Los Arcos G, Vilaró MT, Cortés R, López-Giménez JF, Raurich A, Palacios JM. Chemical neuroanatomy of 5-HT receptor subtypes

in the mammalian brain. In: The Serotonin Receptors: From Molecular Pharmacology to Human Therapeutics. Totowa, N.J: Humana Press; 2006. (The receptors)

[141] Bunin MA, Wightman RM. Paracrine neurotransmission in the CNS: Involvement of 5-HT. Trends in Neurosciences. 1999;**22**(9):377-382

[142] Descarries L, Mechawar N. Ultrastructural evidence for diffuse transmission by monoamine and acetylcholine neurons of the central nervous system. Progress in Brain Research. 2000;**125**:27-47

[143] Freund TF, Buzsáki G. Interneurons of the hippocampus. Hippocampus. 1996;**6**(4):347-470

[144] Varga V, Losonczy A, Zemelman BV, Borhegyi Z, Nyiri G, Domonkos A, et al. Fast synaptic subcortical control of hippocampal circuits. Science. 2009;**326**(5951):449-453

[145] Kosofsky BE, Molliver ME. The serotoninergic innervation of cerebral cortex: Different classes of axon terminals arise from dorsal and median raphe nuclei. Synapse. 1987;**1**(2):153-168

[146] Kiehn O, Rostrup E, Møller M. Monoaminergic systems in the brainstem and spinal cord of the turtle Pseudemys scripta elegans as revealed by antibodies against serotonin and tyrosine hydroxylase. The Journal of Comparative Neurology. 1992;**325**(4):527-547

[147] Alvarez FJ, Pearson JC, Harrington D, Dewey D, Torbeck L, Fyffe RE. Distribution of 5-hydroxytryptamine-immunoreactive boutons on alpha-motoneurons in the lumbar spinal cord of adult cats. The Journal of Comparative Neurology. 1998;**393**(1):69-83

[148] Moukhles H, Bosler O, Bolam JP, Vallée A, Umbriaco D, Geffard M, et al. Quantitative and morphometric data indicate precise cellular interactions between serotonin terminals and postsynaptic targets in rat substantia nigra. Neuroscience. 1997;**76**(4):1159-1171

[149] Ridet JL, Rajaofetra N, Teilhac JR, Geffard M, Privat A. Evidence for nonsynaptic serotonergic and noradrenergic innervation of the rat dorsal horn and possible involvement of neuron-glia interactions. Neuroscience. 1993;**52**(1):143-157

[150] Van Bockstaele EJ, Pickel VM. Ultrastructure of serotonin-immunoreactive terminals in the core and shell of the rat nucleus accumbens: Cellular substrates for interactions with catecholamine afferents. The Journal of Comparative Neurology. 1993;**334**(4):603-617

[151] Chazal G, Ralston HJ 3rd. Serotonin-containing structures in the nucleus raphe dorsalis of the cat: An ultrastructural analysis of dendrites, presynaptic dendrites, and axon terminals. The Journal of Comparative Neurology. 1987;**259**(3):317-329

[152] Kapadia SE, de Lanerolle NC, LaMotte CC. Immunocytochemical and electron microscopic study of serotonin neuronal organization in the dorsal raphe nucleus of the monkey. Neuroscience. 1985;**15**(3):729-746

[153] Liposits Z, Görcs T, Trombitás K. Ultrastructural analysis of central serotoninergic neurons immunolabeled by silver-gold-intensified diaminobenzidine chromogen. Completion of immunocytochemistry with X-ray microanalysis. The Journal of Histochemistry and Cytochemistry. 1985;**33**(6):604-610

[154] Allers KA, Sharp T. Neurochemical and anatomical identification of fast- and slow-firing neurones in the rat dorsal raphe nucleus using juxtacellular labelling methods in vivo. Neuroscience. 2003;**122**(1):193-204

[155] Kocsis B, Varga V, Dahan L, Sik A. Serotonergic neuron diversity: Identification of raphe neurons with discharges time-locked to the hippocampal theta rhythm. Proceedings of the National Academy of Sciences of the United States of America. 2006;**103**(4):1059-1064

[156] Hajós M, Allers KA, Jennings K, Sharp T, Charette G, Sík A, et al. Neurochemical identification of stereotypic burst-firing neurons in the rat dorsal raphe nucleus using juxtacellular labelling methods. The European Journal of Neuroscience. 2007;**25**(1):119-126

[157] Ranade SP, Mainen ZF. Transient firing of dorsal raphe neurons encodes diverse and specific sensory, motor, and reward events. Journal of Neurophysiology. 2009;**102**(5):3026-3037

[158] Pan ZZ, Williams JT. GABA- and glutamate-mediated synaptic potentials in rat dorsal raphe neurons in vitro. Journal of Neurophysiology. 1989;**61**(4):719-726

[159] Pan ZZ, Grudt TJ, Williams JT. Alpha 1-adrenoceptors in rat dorsal raphe neurons: Regulation of two potassium conductances. The Journal of Physiology. 1994;**478**(Pt 3):437-447

[160] Baraban JM, Aghajanian GK. Suppression of firing activity of 5-HT neurons in the dorsal raphe by alpha-adrenoceptor antagonists. Neuropharmacology. 1980;**19**(4):355-363

[161] Davidson C, Stamford JA. Evidence that 5-hydroxytryptamine release in rat dorsal raphé nucleus is controlled by 5-HT1A, 5-HT1B and 5-HT1D autoreceptors. British Journal of Pharmacology. 1995;**114**(6):1107-1109

[162] Stamford JA, Davidson C, McLaughlin DP, Hopwood SE. Control of dorsal raphé 5-HT function by multiple 5-HT(1) autoreceptors: Parallel purposes or pointless plurality? Trends in Neurosciences. 2000;**23**(10):459-465

[163] Starkey SJ, Skingle M. 5-HT1D as well as 5-HT1A autoreceptors modulate 5-HT release in the Guinea-pig dorsal raphé nucleus. Neuropharmacology. 1994;**33**(3-4):393-402

[164] Piñeyro G, de Montigny C, Blier P. 5-HT1D receptors regulate 5-HT release in the rat raphe nuclei. In vivo voltammetry and in vitro superfusion studies. Neuropsychopharmacology. 1995;**13**(3):249-260

[165] Aghajanian GK. Chemical-feedback regulation of serotonin-containing neurons in brain. Annals of the New York Academy of Sciences. 1972;**193**:86-94

[166] Fornal CA, Litto WJ, Metzler CW, Marrosu F, Tada K, Jacobs BL. Single-unit responses of serotonergic dorsal raphe neurons to 5-HT1A agonist and antagonist drug administration in behaving cats. The Journal of Pharmacology and Experimental Therapeutics. 1994;**270**(3):1345-1358

[167] Sari Y, Miquel MC, Brisorgueil MJ, Ruiz G, Doucet E, Hamon M, et al. Cellular and subcellular localization of 5-hydroxytryptamine1B receptors in the rat central nervous system: Immunocytochemical, autoradiographic and lesion studies. Neuroscience. 1999;**88**(3):899-915

[168] Riad M, Garcia S, Watkins KC, Jodoin N, Doucet E, Langlois X, et al.

Somatodendritic localization of 5-HT1A and preterminal axonal localization of 5-HT1B serotonin receptors in adult rat brain. The Journal of Comparative Neurology. 2000;**417**(2):181-194

[169] Pan ZZ, Colmers WF, Williams JT. 5-HT-mediated synaptic potentials in the dorsal raphe nucleus: Interactions with excitatory amino acid and GABA neurotransmission. Journal of Neurophysiology. 1989;**62**(2):481-486

[170] Wang RY, Aghajanian GK. Antidromically identified serotonergic neurons in the rat midbrain raphe: Evidence for collateral inhibition. Brain Research. 1977;**132**(1):186-193

[171] Wang RY, Aghajanian GK. Collateral inhibition of serotonergic neurones in the rat dorsal raphe nucleus: Pharmacological evidence. Neuropharmacology. 1978;**17**(10):819-825

[172] García-Ávila M, Torres X, Cercós MG, Trueta C. Specific localization of an auto-inhibition mechanism at presynaptic terminals of identified serotonergic neurons. Neuroscience. 2021;**458**:120-132

[173] Cercós MG, De-Miguel FF, Trueta C. Real-time measurements of synaptic autoinhibition produced by serotonin release in cultured leech neurons. Journal of Neurophysiology. 2009;**102**(2):1075-1085

[174] Piñeyro G, Blier P. Autoregulation of serotonin neurons: Role in antidepressant drug action. Pharmacological Reviews. 1999;**51**(3):533-591

[175] Vizi ES, Fekete A, Karoly R, Mike A. Non-synaptic receptors and transporters involved in brain functions and targets of drug treatment. British Journal of Pharmacology. 2010;**160**(4):785-809

[176] Jennings KA, Lesch K, Sharp T, Cragg SJ. Non-linear relationship between 5-HT transporter gene expression and frequency sensitivity of 5-HT signals. Journal of Neurochemistry. 2010;**115**(4):965-973

[177] Hirasawa H, Puopolo M, Raviola E. Extrasynaptic release of GABA by retinal dopaminergic neurons. Journal of Neurophysiology. 2009;**102**(1):146-158

[178] Puopolo M, Hochstetler SE, Gustincich S, Wightman RM, Raviola E. Extrasynaptic release of dopamine in a retinal neuron: Activity dependence and transmitter modulation. Neuron. 2001;**30**(1):211-225

[179] Trueta C, De-Miguel FF. Extrasynaptic exocytosis and its mechanisms: A source of molecules mediating volume transmission in the nervous system. Frontiers in Physiology. 2012;**3**:319

[180] Bruns D, Riedel D, Klingauf J, Jahn R. Quantal release of serotonin. Neuron. 2000;**28**(1):205-220

[181] Agnati LF, Leo G, Zanardi A, Genedani S, Rivera A, Fuxe K, et al. Volume transmission and wiring transmission from cellular to molecular networks: History and perspectives. Acta Physiologica (Oxford, England). 2006;**187**(1-2):329-344

[182] Henderson LP. The role of 5-hydroxytryptamine as a transmitter between identified leech neurones in culture. Journal of Physiology (London). 1983;**339**:309-324

[183] Kuffler DP, Nicholls J, Drapeau P. Transmitter localization and vesicle turnover at a serotoninergic synapse between identified leech neurons in culture. The Journal of Comparative Neurology. 1987;**256**(4):516-526

[184] Nicholls JG, Kuffler DP. Quantal release of serotonin from presynaptic nerve terminals. Neurochemistry International. 1990;**17**(2):157-163

[185] del Castillo J, Katz B. Quantal components of the end-plate potential. Journal of Physiology (London). 1954;**124**(3):560-573

[186] Llinás R, Sugimori M, Silver RB. The concept of calcium concentration microdomains in synaptic transmission. Neuropharmacology. 1995;**34**(11):1443-1451

[187] Stewart RR, Adams WB, Nicholls JG. Presynaptic calcium currents and facilitation of serotonin release at synapses between cultured leech neurones. The Journal of Experimental Biology. 1989;**144**:1-12

[188] Trueta C, De-Miguel FF. Extrasynaptic exocytosis and its mechanisms: A source of molecules mediating volume transmission in the nervous system. Frontiers in Physiology. 2012;**3**:3. Available from: http://www.ncbi.nlm.nih.gov/pmc/articles/PMC3432928/

[189] Bunin MA, Wightman RM. Quantitative evaluation of 5-hydroxytryptamine (serotonin) neuronal release and uptake: An investigation of extrasynaptic transmission. The Journal of Neuroscience. 1998;**18**(13):4854-4860

[190] Mansari ME, Wiborg O, Mnie-Filali O, Benturquia N, Sánchez C, Haddjeri N. Allosteric modulation of the effect of escitalopram, paroxetine and fluoxetine: In-vitro and in-vivo studies. The International Journal of Neuropsychopharmacology. 2007;**10**(1):31-40

[191] Pudovkina OL, Cremers TIFH, Westerink BHC. Regulation of the release of serotonin in the dorsal raphe nucleus by alpha1 and alpha2 adrenoceptors. Synapse. 2003;**50**(1):77-82

[192] Swanson CJ, Blackburn TP, Zhang X, Zheng K, Xu ZQD, Hökfelt T, et al. Anxiolytic- and antidepressant-like profiles of the galanin-3 receptor (Gal3) antagonists SNAP 37889 and SNAP 398299. Proceedings of the National Academy of Sciences of the United States of America. 2005;**102**(48):17489-17494

[193] Wright IK, Upton N, Marsden CA. Effect of established and putative anxiolytics on extracellular 5-HT and 5-HIAA in the ventral hippocampus of rats during behaviour on the elevated X-maze. Psychopharmacology. 1992;**109**(3):338-346

[194] Marinesco S, Wickremasinghe N, Carew TJ. Regulation of behavioral and synaptic plasticity by serotonin release within local modulatory fields in the CNS of Aplysia. The Journal of Neuroscience. 2006;**26**(49):12682-12693

[195] Hentall ID, Pinzon A, Noga BR. Spatial and temporal patterns of serotonin release in the rat's lumbar spinal cord following electrical stimulation of the nucleus raphe magnus. Neuroscience. 2006;**142**(3):893-903

[196] Kia HK, Miquel MC, Brisorgueil MJ, Daval G, Riad M, El Mestikawy S, et al. Immunocytochemical localization of serotonin1A receptors in the rat central nervous system. The Journal of Comparative Neurology. 1996;**365**(2):289-305

[197] Zhou FC, Tao-Cheng JH, Segu L, Patel T, Wang Y. Serotonin transporters are located on the axons beyond the synaptic junctions: Anatomical and functional evidence. Brain Research. 1998;**805**(1-2):241-254

[198] Vizi ES, Kiss JP, Lendvai B. Nonsynaptic communication in the central nervous system. Neurochemistry International. 2004;**45**(4):443-451

[199] Trueta C, Méndez B, De-Miguel FF. Somatic exocytosis of serotonin mediated by L-type calcium channels in cultured leech neurones. Journal of Physiology (London). 2003;**547**(Pt 2):405-416

[200] Leon-Pinzon C, Cercós MG, Noguez P, Trueta C, De-Miguel FF. Exocytosis of serotonin from the neuronal soma is sustained by a serotonin and calcium-dependent feedback loop. Frontiers in Cellular Neuroscience. 2014;**8**:169

[201] Trueta C, Sánchez-Armass S, Morales MA, De-Miguel FF. Calcium-induced calcium release contributes to somatic secretion of serotonin in leech Retzius neurons. Journal of Neurobiology. 2004;**61**(3):309-316

[202] Kaushalya SK, Desai R, Arumugam S, Ghosh H, Balaji J, Maiti S. Three-photon microscopy shows that somatic release can be a quantitatively significant component of serotonergic neurotransmission in the mammalian brain. Journal of Neuroscience Research. 2008;**86**(15):3469-3480

[203] Sarkar B, Das AK, Arumugam S, Kaushalya SK, Bandyopadhyay A, Balaji J, et al. The dynamics of somatic exocytosis in monoaminergic neurons. Frontiers in Physiology [Internet]. 2012;**3**:3. Available from: http://www.frontiersin.org/Membrane_Physiology_and_Biophysics/10.3389/fphys.2012.00414/abstract

[204] Kaushalya SK, Nag S, Ghosh H, Arumugam S, Maiti S. A high-resolution large area serotonin map of a live rat brain section. Neuroreport. 2008;**19**(7):717-721

[205] Colgan LA, Cavolo SL, Commons KG, Levitan ES. Action potential-independent and pharmacologically unique vesicular serotonin release from dendrites. The Journal of Neuroscience. 2012;**32**(45):15737-15746